Contents

List of figures

List of tables

List of boxes

Acknowledgements

This manual is the result of the combined efforts of many staff of Oxfam GB. Part 1 is largely based on a review of cash-transfer programmes as an alternative to food aid, conducted by Hisham Khogali (Food and Nutrition Adviser) in 2001. Chapter 4, on implementing cash-for-work programmes, is based on guidelines produced by Sarah Lumsden (Programme Manager) and Emma Naylor (Humanitarian Programme Co-ordinator) in 2002, drawing on the experiences of the Oxfam GB team in Kenya in Turkana and Wajir. In 2003, Frances Mason (Food and Nutrition Adviser) did an initial edit of the guidelines. Pantaleo Creti (Emergency Food Security and Livelihood Adviser) wrote Chapters 3 and 5 on cash grants and vouchers. Pantaleo Creti and Susanne Jaspars (Senior Emergency Food Security and Livelihood Adviser until June 2005) compiled the latest version of the guidelines. Key comments on case studies were contributed by Patrick Alix (Humanitarian Programme Co-ordinator, Haiti), Sophia Dunn, Silke Pietszch, Paola Siragusa, and Jonathan Brass (Emergency Food Security and Livelihoods Humanitarian Support Personnel), Lilian Mutiso (Food Security Adviser, Zimbabwe), Ann Witteveen (Regional Food Security Co-ordinator for Southern Africa), Vivien Lee (Consultant), Michael Schultz (Livelihoods Co-ordinator, Aceh), and Yves Reynaud (Food Security Co-ordinator, Somaliland). Laura Phelps and Chris Leather (Emergency Food Security and Livelihood Advisers) made significant contributions to the final version.

We thank Paul Harvey for his extensive and detailed comments on the trial version of this guide. We also thank Lesley Adams, Tilleke Kiewied, Sonya Lejeune, Jacqueline Frize, Robin Palmer, Niall Cassidy, Simon Narbeth, Mathias Rickli, Ines Smyth, Charles Antoine-Hoffman, Lisa Ernoul, and Adeola Akintoye for their comments.

Introduction

The aim of this practical guide is to support the implementation of cash programmes in emergencies. It is based on the experience of Oxfam GB over five years (2000–2005) in a variety of disaster contexts.

Oxfam GB (referred to from now on as 'Oxfam') has used cash interventions as part of its response to the needs of communities affected by droughts, floods, hurricanes, and cyclones, and the needs of displaced people and people experiencing chronic food insecurity as a result of protracted conflict and/or poverty. This guide makes extensive reference to responses to the tsunami that struck the Indian Ocean region in December 2004. Most of Oxfam's experience relates to cash-for-work programmes, but in the past three years Oxfam staff have increasingly implemented cash grants and voucher programmes. Many other agencies are implementing cash programmes; when possible and appropriate, we have drawn on their materials to inform these guidelines. However, this book is mainly based on Oxfam's experience.

All cash programmes have the following broad aim: *to increase the purchasing power of disaster-affected people to enable them to meet their minimum needs for food and non-food items; or to assist in the recovery of people's livelihoods.*[1]

In reality, food aid dominates emergency response. However, food aid, as a resource transfer, is sometimes highly inefficient. It is not always the right response, even when the disaster-affected population are unable to meet their immediate food needs. Oxfam's guiding principles for response to food crises, produced in November 2002, promote alternatives to food aid where appropriate and feasible. The alternatives include cash vouchers and food vouchers, cash-for-work programmes, cash grants, market support, and production support (for agriculture and livestock). According to the Sphere Minimum Standards for Disaster Response, in a guidance note on the first food-security standard:

General food distribution may not be appropriate when

Adequate supplies of food are available in the area (and the need is to address obstacles to access).

A localised lack of food availability can be addressed by the support of market systems.[2]

Cash-transfer interventions are increasingly considered by donors and humanitarian agencies as an appropriate emergency response to meet immediate needs for food and non-food items, and to support the recovery of livelihoods. Cash interventions can be used to meet any need for which there is a private market. The cash transfers described in this book are intended to enable recipients to obtain goods and services directly from local traders and service providers, rather than from an aid agency. The aid agency is not directly involved in the procurement, transportation, or provision of goods and services. Cash transfers often therefore meet people's needs more quickly than commodity distribution, because they reduce the logistics involved. At the same time, they stimulate the local economy. Moreover, cash transfers are more dignified than in-kind distributions (of items such as food aid, jerry cans, cooking stoves, seeds, and tools), because they give disaster-affected populations the option of spending according to their own priorities.

This book is intended to help programme managers and technical specialists to decide whether or not cash interventions would be appropriate in particular circumstances. It also offers guidance on the practical implementation of cash grants, cash-for-work, and voucher programmes. It does not explain how to do an emergency needs assessment. Nor does it cover other types of cash-transfer intervention: micro-finance, credit, insurance, tax breaks, or budget support, for example.

Table 1 lists basic definitions of the most common types of cash programme in emergencies.

Table 1: Forms of cash transfer

Cash grants	The provision of money to targeted households, either as emergency relief to meet their basic needs for food and non-food items, or as grants to buy assets essential for the recovery of their livelihoods. Cash grants for livelihood recovery differ from micro-finance in that beneficiaries are not expected to repay the grants, and the financial services provided are not expected to continue in the long term. Both cash grants and micro-finance may be accompanied by training to upgrade the recipients' skills.
Cash for work	Payment for work on public or community works programmes. The cash wages help people to meet their basic needs, and the community project helps to improve or rehabilitate community services or infrastructure. Cash for work differs from casual labour in that it is targeted at the poorest or most food-insecure members of the community.
Vouchers	Vouchers provide access to pre-defined commodities. They can be exchanged in a special shop or from traders in fairs and markets. The vouchers may have either a cash value or a commodity value. Vouchers have been most commonly used for the provision of seeds and livestock, but they can also be used to provide food.

This guide is written for a range of technical staff and managers who are actively involved in the implementation of cash interventions:

- *food-security specialists and programme managers,* who identify needs, assess whether cash programming is appropriate and feasible, and plan or implement monitoring and evaluation;
- *public-health engineers* engaged in cash-for-work programming, if the work projects are related to water and sanitation: for example, the digging of drainage channels, the disposal of solid waste, and the construction of latrines;
- *public-health promoters,* who in such cases are responsible for identifying the projects, and who mobilise the community to select the most vulnerable groups for inclusion in cash programmes;
- *logisticians,* who identify and source the materials for all cash-for-work programmes;
- *finance officers or accountants,* who are responsible for handing out money and recording distributions.

The guide is divided into two parts. Part 1 considers the rationale for cash interventions and the details of the decision-making process. Part 2 discusses the practical implementation of various cash-transfer programmes.

Chapter 1 provides the theoretical basis for cash transfers and presents some of their advantages. Many agency staff and local governments are reluctant to implement cash programmes because of the assumed risks, so the final part of this chapter addresses each of the commonest fears in turn.

Chapter 2 covers the types of information that are needed in order to determine when a cash intervention is appropriate – including in particular the market information that is necessary for the planning of cash programmes. The final part of this chapter offers some criteria for determining the most appropriate response: cash grants, vouchers, or a cash-for-work programme.

Chapter 3 provides information on how to implement cash grants to meet the basic needs of the poorest, or those worst affected by a disaster. It explains how to implement cash grants to support livelihood recovery. It also covers some elements common to all cash programmes: for example, the establishment of community-based relief committees and community-based targeting. It also includes a section on monitoring and evaluation, which is common to all cash interventions.

Chapter 4 describes the implementation of cash-for-work programmes. It presents all the steps necessary to implement them, such as identifying the projects, setting wage rates, and making the cash payments, as well as aspects of monitoring and evaluation that are specific to cash-for-work programmes. This chapter makes extensive reference to examples from Oxfam's own experience.

Chapter 5 explains how to implement voucher schemes, particularly via fairs and shops. It describes all the necessary steps to implement such programmes, and suggests some specific indicators for monitoring voucher programmes.

The book concludes with some reflections on the role of cash interventions in linking emergencies with longer-term programming, and the main challenges faced by humanitarian and development organisations in the coming years.

The appendices provide examples of documents such as monitoring forms, and a logical framework.

These guidelines will be updated after one year. Meanwhile Oxfam is conducting field-based research, both independently and in conjunction with the Overseas Development Institute (ODI) and the World Food Programme (WFP), to define different types of cash intervention and to identify impact criteria. The lessons from these initiatives will inform the updated guidelines.

We would value readers' contributions to the update of the guidelines. You can send your comments to cash_guidelines@oxfam.org.uk. We would also appreciate receiving information about any cash programmes that readers are implementing, including assessment, monitoring, and evaluation reports.

Part 1 | Planning cash-transfer programmes

1 | Why provide cash as a response to emergencies?

The rationale for cash interventions

Nowadays almost everyone lives in a cash economy: people earn wages, sell goods or services, and buy what they need with cash. Giving people money is therefore the most obvious and simple way of providing assistance in emergencies. But emergency relief is dominated by the distribution of in-kind commodities, in particular food aid. In many emergencies, the problem is that people are unable to buy food and other basic goods – not that such items are unavailable. If markets are still functioning, emergency-affected populations can be supported to buy the commodities that they need on the market.

Support for the idea of cash interventions is derived from Amartya Sen's 'entitlement theory', from studies of people's coping strategies in response to emergencies, and from experience of the range of livelihood needs that arise following a disaster.

Entitlement theory states that famines are often caused not by lack of food, but by individuals' inability to get access to whatever food exists.[1] Entitlement failure could occur through loss of income or loss of employment, or high food prices, or reduced food availability. Famines or food insecurity are therefore as much a result of people's inability to buy food as they are caused by a decline in overall food availability or food production. It is therefore logical to conclude that if famine results from a lack of purchasing power, it can be addressed through income transfers.

People affected by disaster or famine often seek an income, for example by moving to another area to find work ('labour migration') or by selling off their assets. They must balance the need to maintain their current food consumption against the need to protect their future income-generating capacity and livelihoods. Providing cash to populations affected by famine or disaster may help them to avoid resorting to coping strategies that are damaging to their livelihoods or dignity, such as the sale of productive assets, or sex work, or illegal or violent activities.

Disasters may affect several aspects of people's livelihoods, their capabilities, assets, and activities required for a means of living. Needs may range from essential livelihood assets (such as agricultural inputs, livestock, tools, and raw materials), to a range of food commodities, and to non-food needs such as kitchen utensils, hygiene items, and clothes. Oxfam considers livelihoods in terms of both food security and income security. Moreover, the impact of sudden-onset disasters, such as floods, hurricanes, cyclones, and the Indian Ocean tsunami of 2004, extends beyond livelihoods: shelter, health services, and education are also affected.

Cash transfers as a form of famine relief and disaster relief are not new. In 1948, the British colonial administration in Sudan distributed cash, coffee, and train tickets to famine-affected populations. In Bangladesh there is a long history of cash relief. Many developed countries provide cash transfers as part of their social welfare systems.

Advantages of cash transfers

The experience of Oxfam and others shows that cash-based programmes, in appropriate circumstances, are less costly and better adjusted to people's needs and preferences than the distribution of commodities in kind. And they can be more timely. The advantages of cash interventions are summarised in Box 1 on page 8.

Oxfam and other humanitarian agencies have in the past considered using cash interventions, as an alternative or a complement to food aid. Food aid is most commonly supplied from donor countries, which means that the commodities are not necessarily appropriate to the culture of the recipients, and it may take 4–5 months to arrive in the disaster-affected area, by which time it is too late to meet immediate needs. Food aid which arrives late or is delivered when there is no actual food shortage may adversely affect local markets, reduce food prices, and therefore risk increasing the vulnerability of food producers and traders.

The supply of seeds provides another example where cash or vouchers would be a more appropriate and effective response than distributions in kind, in particular when seed is available in sufficient quantity within a reasonable distance of the target area. Seed vouchers which have a cash value and can be exchanged in local fairs give farmers greater choice, strengthen local procurement systems, and often are more timely and cost-effective than distributions of improved varieties and certified seeds.[2]

In Oxfam's experience, the advantages of cash interventions far outnumber the potential disadvantages (which might more accurately be described as fears). These are discussed in the next section.

Oxfam's experience shows that most of these fears have not been borne out in practice, or that they can be successfully managed.

Box 1: Advantages of cash transfers

Depending on the circumstances, cash transfers may offer the following advantages:

Choice: cash gives households a greater degree of choice and permits them to spend money according to their own priorities.

Cost-effectiveness: cash is likely to be cheaper and faster to distribute than alternatives such as restocking, seed distribution, and food distribution.

Dignity: offering cash maintains people's dignity, by giving them choice. Delivery mechanisms do not treat them as passive recipients of relief.

Economic recovery: injections of cash have potential benefits for local markets and trade.

Flexibility: cash can be spent on both food and non-food items and is easily invested in livelihood security.

Empowerment: cash can improve the status of women and marginalised groups.

During monitoring and evaluations of Oxfam programmes in Bangladesh, Cambodia, Kenya, Uganda, Afghanistan, and Haiti,[3] recipients stated that they preferred cash-based programmes to commodity-based assistance because cash gave them choices: to buy goods and services according to their own priorities, to meet immediate needs, and to invest in future livelihood assets. When cash is used to buy food, people can buy the familiar foods that they like.

The ways in which project beneficiaries may spend cash distributed by aid agencies are summarised in Box 2.

The nature of people's expenditure varies according to the context, including other types of relief distributed at the same time, the method of payment, the quantity of cash distributed, and the timing of payment in relation to the seasonal calendar. In Oxfam programmes in Uganda, Afghanistan, and Haiti, beneficiary populations who received cash spent most of it on food.[4] In Afghanistan they spent up to 90 per cent of the cash received on food. The remainder was often spent on clothes and medicine, with a few households being able to invest in livestock or pay off debts. When people are receiving cash in addition to food aid, then cash is less likely to be spent on food. For example, when cash was given in addition to food aid in Turkana, Kenya, 81 per cent of the money distributed was spent

Box 2: Examples of beneficiaries' use of cash

Purchase of food, kitchen utensils, clothes

Paying off debts and loans. Extending credit.

Payment of school costs: fees, clothes, transport

Purchase of livestock and agricultural inputs

Payment for health care

Setting up small shops

Purchase of tools for petty trade: for example, wood cutting, donkey carting

on livelihood recovery (including re-stocking, business inputs, and school fees).[5] In Indonesia, where people were receiving food rations at the same time as cash, day-to-day expenditure included snacks, cigarettes, fish, vegetables, sugar, and coffee, while one-off larger expenditures included community contributions (for example, for religious festivals), clothes, and gold (as a form of saving). Although expenditure on items such as cigarettes and coffee might not be considered important for household food security, freedom to spend money on these items was seen as a significant step towards restoration of 'normality'.[6] Oxfam in general is opposed to smoking, but we believe that switching to in-kind assistance in such situations would not prevent people smoking. In-kind assistance would release income that would otherwise be spent on those commodities, so increasing available income to buy cigarettes.

Small regular payments are more likely to be used to buy food, whereas larger lump sums are more likely to be spent on productive assets and re-establishing economic activities. In Turkana, Kenya, small cash transfers were used for buying foodstuffs not included in the relief ration, for paying off debts, and for partial payment of school fees. Where cash was paid in a lump sum, it was spent on productive assets such as goats, setting up small shops, tools for firewood cutting, and donkey carting.[7] In Ethiopia, an evaluation by Save the Children UK found that when cash payments exceeded minimum needs, and timing coincided with critical times in the seasonal calendar, then households could make strategic investments, for example by re-negotiating contractual agreements for sharecropping, and purchasing small stock or plough oxen.[8] In general, the larger the payment, the more likely this will be spent on livelihood recovery.

In some societies, men and women spend money differently. In Bangladesh, an evaluation conducted by Khogali in 2001 found that women often made joint decisions with men about expenditure, but they also retained some of the cash for future unforeseen expenditure. In general, women gave more thought to future needs, investing in productive asset creation, paying off loans, and saving. Men tended to keep the money they earned, but gave money to women for specific purchases. Men appeared to save less than women, spending money mainly on paying off loans, and buying food and clothes. In many other contexts, there was no difference in the expenditure patterns of men and women.

In some programmes women's status was improved, both in the household and in the community, by their ability to earn and control income.

Few evaluations have been conducted to estimate the cost-effectiveness of cash transfers as opposed to commodity distributions. An evaluation of a cash-for-work (CFW) scheme in Kenya, which compared food distribution, cash-for-work, and livestock restocking interventions, found that 'cash for work was the most cost-effective recovery intervention in terms of the cost of providing for the subsistence of beneficiaries, without even considering the value of the work undertaken'.[9] The sourcing of food aid in Western donor countries is a very inefficient way of meeting food needs. For example, a study in Ethiopia found that cash transfers were 6–7 per cent cheaper than local food purchase, and between 39 and 46 per cent cheaper than imported relief food.[10] Similarly, in Democratic Republic of Congo, it cost $15 to deliver an amount of imported food aid which could have been purchased on the local market for $1.[11] This inefficiency increases when beneficiaries use food relief as a resource to meet other household needs – that is, when they sell their food relief to buy other food items, to pay for health care and education, or to meet other essential needs.

Cash transfers can stimulate economic recovery by encouraging traders to move supplies from areas of food surplus to areas of food deficit. This helps to maintain prices (and production) in areas of surplus. Experience has shown that injections of cash have boosted trade in the following ways:

- A significant proportion of the cash transfer was invested in trade.
- Money was frequently used as capital to set up small businesses such as kiosks, teashops, and other small market enterprises.
- Cash transfers boosted purchases from local traders.
- Most of the livestock purchased was obtained from local producers. [12]

The CFW projects themselves, like food-for-work projects, often have an impact on both food security and public health. Projects such as clearance and rehabilitation of roads facilitate trade into the area, and stimulate travel to markets outside the affected area. Agricultural rehabilitation through de-silting, bunding, training, and tree planting should result in increased food production. Dam construction, well cleaning, canal clearing, and rainwater management improve water supplies for humans and livestock.

Addressing fears about cash distribution

A number of fears about distributing cash deter humanitarian agencies and donors from implementing cash programmes more frequently, even when assessments have shown that it would be the most appropriate intervention. Such fears include the following:

- Cash is difficult to target, because everyone wants money.
- Cash injections may cause inflation, which means that those not included in the programme will suffer.
- Cash transfers may increase security risks, either for the agency or for the beneficiaries.
- Women may not have control over the income, so it will not be spent on household needs. If women receive money, this could provoke family disputes or domestic violence. *⌐ little evidence*
- It may be spent on the 'wrong' things, such as tobacco, alcohol, or drugs. *towards*
- Cash may be diverted from its intended targets, because it is attractive to powerful members of the community, as well as to the most vulnerable.
- Direct cash transfers will undermine development programmes such as micro-finance.
- NGO-implemented cash-transfer programmes will set up parallel systems and undermine government social-welfare systems, or remove government responsibility to provide them.
- Some or all of the money will be seized by landowners or unscrupulous lenders.

Such concerns are often raised prior to cash-intervention programmes in countries where there is no previous experience of them. Following five years of field experience and six evaluations[13] of cash-transaction programmes, Oxfam has concluded that many of the perceived risks and fears are not borne out in practice. This does not mean that they should not be taken into consideration when programmes are designed, monitored, and evaluated, but they should not be the primary reason why a cash-transaction programme is not implemented in the first place. Security is a very valid concern and it should always be managed as such.

Table 2 on page 12 addresses many of these fears in a question-and-answer format, using Oxfam experience to answer each question.

Table 2: Fears associated with cash-distribution programmes

Questions	Answers
Is targeting more difficult because cash is of value to everyone?	When targeting cash, Oxfam often uses a community-based approach similar to that used effectively when targeting food aid. Communities are involved in the selection of beneficiaries and the management of the programme. Alternative targeting methods include self-targeting by setting wages slightly under the minimum wage (in the case of cash for work), or targeting on the basis of clearly identifiable criteria, for example *destruction of house*.
Do large injections of cash cause inflation and increase local food prices?	Even the largest payments that Oxfam has made represent only a small part of the local economy. Most evaluations show that the cash programme had no inflationary effect. Where an increase in food prices is considered to be a risk, food aid can be distributed alongside the cash. In Uganda, beneficiaries experienced some price increases in local village shops, but villagers overcame this by searching out better-value commodities in larger trading centres. Inflation might be a problem in larger-scale programmes, and the impact of large amounts of cash on the economy needs to be closely monitored.
Could cash distributions threaten the security of the implementing agency and targeted beneficiaries?	Any distribution of resources entails security risks, and there is no evidence that cash distributions create greater risks than in-kind distributions. The risks associated with different forms of distribution need to be considered carefully, and efforts should be made to minimise them. Recommendations for doing this are given in Chapter 4. In Uganda, beneficiaries reported that they spent money immediately after receiving it, because they feared losing cash to raids and theft.
Does the provision of cash to women provoke social problems such as family disputes and domestic violence?	When distributing cash grants and vouchers, Oxfam actively promotes the targeting of women as heads of household. Oxfam has found that when women are the direct recipients of cash transfers, they can gain a greater share of household income, which increases their status

Questions	Answers
	within communities and gives them greater decision-making authority within households. Cash for work can promote gender equity by payment of equal wages to women and men.
	Even in societies where gender roles are very strictly divided and women do not normally do paid work, women's participation in CFW projects has been accepted. In some societies initial difficulties were encountered, but they were overcome by community-sensitisation work; family disputes were settled by village committees.
Is cash likely to be used to buy non-essential items like tobacco and alcohol?	Oxfam's monitoring and evaluation shows that beneficiaries of cash-transfer programmes use the cash mainly for food purchase, repayment of loans, school books/fees/uniforms, clothes, livestock, and agricultural inputs. In some programmes some cash was spent on cigarettes and other items considered non-essential in terms of nutrition or livelihoods. Oxfam believes that the same risk exists with in-kind distribution, and that stopping cash distributions will not stop people buying non-essential commodities.
Is cash more likely to be diverted by elites and authorities because it is more attractive than commodities?	There is certainly a risk of this, and the main way to prevent it is by rigorous monitoring. Similar risks exist for commodity distributions. In all cases, the risks associated with cash transfers should be compared with the risks associated with commodity distributions. In some cases, cash for work has been shown to be less prone to diversion than food aid, since people feel a greater sense of ownership over money that they have earned by working. They will therefore insist on receiving the payment agreed for work done.
Will cash transactions undermine development programmes such as micro-finance schemes?	In certain circumstances where people have lost everything, for example natural disasters, conflict, and displacement, experience shows that the most vulnerable households are unwilling or unable to obtain loans that they are unlikely to be able to repay. Cash transactions

continued ...

Questions	Answers
	can meet immediate needs and rehabilitate livelihoods in the short term, and such programmes can become integrated into longer-term soft micro-finance (such as micro-credit, rotational funds, and insurance) or income-generation programmes.
Will the cash transaction set up a parallel system?	After some disasters, the government may distribute cash grants or vouchers. It is important that humanitarian agencies do not undermine these by creating a parallel system. However, in contexts where local government support systems are not targeted, or are dysfunctional due to corruption or low capacity, the agency might decide to run a carefully monitored and targeted cash-transaction programme.
Will private lenders to whom the community is in debt seize the money?	If the community is heavily indebted to moneylenders, middlemen, landowners, or warlords, it may be necessary for implementing agencies to negotiate payment holidays, or reduced interest rates, on their behalf. In many cases, however, repayment of debts will be the beneficiary population's highest priority, and the size of the cash transfer should take this into account.

2 | When is a cash-intervention strategy appropriate?

Cash or commodity distribution?

An emergency assessment should determine the impact of the emergency on people's life, health, and dignity. This includes an assessment of their access to water, sanitation, and health care; their need for protection; and a food-security assessment which considers both food and income security. The minimum necessary is a food-security assessment, which in addition to assessing people's ability to meet their food needs considers the impact of changes in income on their ability to obtain water and health care, and to meet other basic needs. For Oxfam staff, more information about the agency's approach to emergency assessments is available in the Oxfam GB Emergency Response Manual (2005); for information about food-security assessment and response, staff are referred to 'Oxfam Guidelines on Emergency Food Security Assessment and Response' (2003).

Cash interventions are appropriate in the following circumstances:

- before the emergency people used to purchase a significant proportion of essential goods and services through market mechanisms;

- a shock[1] has resulted in a decline in people's sources of food and income, which means they can no longer meet their basic needs and/or are adopting coping strategies which are damaging to their livelihoods or dignity;

- sufficient food supplies and/or other goods are available locally to meet the needs;

- markets are functioning and accessible;

- cash can be delivered safely and effectively.

Many emergency contexts will meet these criteria. Exceptions include, for example, situations where roads and bridges are destroyed as a result of floods, hurricanes, or cyclones, hindering people's access to markets. Even in these situations, however, cash may be appropriate after the first few weeks, when markets are accessible again. In some protracted conflicts, markets are inaccessible or they are targeted by the warring parties. In this case, in-kind assistance would be the appropriate response to meet food needs. Cash alone is not appropriate for the rehabilitation of malnourished children, since it requires specialised medical and nutritional care.

Table 3 lists key questions to ask to determine whether cash interventions are appropriate.

Table 3: Questions to answer and methodologies to apply when assessing the appropriateness of a cash intervention

Issue	Key questions	Methods
Needs	What was the impact of the shock on people's food and income sources?	Participatory approaches Interviews, surveys
	What was the impact of the shock on people's assets, in particular those essential to their livelihoods?	
	Are people able to meet their basic needs with the food and income available after the shock?	
	Are people able to recover their livelihoods with the assets and income available after the shock?	
	What strategies are people using to cope with food insecurity or income insecurity? What impact do the strategies have on livelihoods and dignity?	
	What are people likely to spend cash on?	
	Do emergency-affected populations have a preference for cash or in-kind approaches?	

Issue	Key questions	Methods
Social relations and power within the household and community	Do men and women have different priorities? How is control over resources managed within households? What are the differences within the community in terms of control over resources? What impact will cash distributions have on existing social and political divisions?	Separate interviews with men and women Ensure that the different social, ethnic, political, and wealth groups are included in interviews
Food availability	Is food available nationally and locally in sufficient quantities and quality? Will the normal seasonal fluctuations affect food availability? Will government policy or other factors affect food availability?	Interviews and focus-group discussions with producers National and local statistics Agricultural calendars Government subsidies and policies
Markets	Are markets in the affected area operating and accessible? Are essential basic items available in sufficient quantities and at reasonable prices? Are there any restrictions on the movement of goods? Is the market competitive, i.e. is the number of suppliers large enough in relation to the number of buyers? Is the market integrated, i.e. are market services functioning and enabling goods to move from areas of surplus to areas of deficit?	Interviews and focus-group discussions with traders Price monitoring in key markets Interviews and focus-group discussions with moneylenders, debtors, and creditors Assess the volume of cash being provided by the project, compared with other inflows such as remittances Ensure that remote areas are covered when analysing how markets work Oxfam Market Analysis Tool (see Figure 2 on page 23)

continued ...

Issue	Key questions	Methods
	Are traders able and willing to respond to an increase in demand?	
	What are the risks that cash will cause inflation in prices of key products?	
Security and delivery mechanisms	What are the options for delivering cash to people?	Mapping of financial transfer mechanisms
	Are banking systems or informal financial transfer mechanisms functioning?	Interviews with banks, post offices, remittance companies
	What are the risks of cash benefits being taxed or seized by elites or warring parties?	Interviews with potential beneficiaries about local perceptions of security and ways of transporting, storing, and spending money safely
	How do these risks compare with the risks posed by in-kind alternatives to cash?	Analysis of the risks of moving or distributing cash
		Analysis of the political-economic context
Corruption	What are the risks of cash being diverted by local elites or project staff?	Assessment of existing levels of corruption and diversion
	How do these compare with the risks of providing in-kind alternatives?	
	What accountability safeguards are available to minimise these risks?	

(Adapted with permission from Harvey 2005)

Market assessments

A market assessment is essential in order to determine whether a cash intervention is appropriate in any particular situation. At the very least it should establish whether markets are functioning or likely to recover quickly following a disaster, and whether the basic items that people need are available in the market. This can be done quickly in the first few days following a rapid-onset emergency, by visiting markets and interviewing traders. When time permits, a more detailed market assessment should be conducted, as part of an in-depth livelihoods analysis. Such an in-depth analysis is recommended as part of emergency-preparedness measures.

In an emergency assessment, seven basic questions need to be answered in order to determine whether cash transfer is the most appropriate response (see also Table 3 on page 16). They all relate to the fundamental question: *will an increase in demand for basic goods, created by a distribution of cash, be met by the market?* The questions, listed below, should be answered for each of the goods for which a need has been identified: for example, staple foods, vegetables, non-food items (cooking pots, stoves), and livelihood assets.

Q1. Are markets operating and accessible?

The first thing to check is whether markets in the emergency-affected areas are actually operating. Even if local markets are not operating, check whether the affected population has easy access to markets that are operating, or if traders can easily supply local markets from others nearby if there is a demand for goods.

The population's physical access to a market is determined by its location, and the time and expense involved in accessing it; the frequency of transport to the market; and the number of months in a year when market access is limited by snow, floods, conflict, etc.

Q2. Are the basic items that people need available on the market in sufficient quantities and at reasonable prices?

If an assessment finds that people are not able to meet their basic needs or to protect or recover their livelihoods without assistance, then the next step is to find out the availability and market price of the items that people need. The items that people need must be available, or potentially available, through markets and traders, for cash to be an appropriate intervention to meet those needs.

Q3. Are there restrictions on the movement of goods?

When assessing the feasibility of cash interventions, it is important to discover whether government policies restrict movements of goods from one part of the country to another, or restrict imports from or exports to other countries. During food crises it is common for countries to restrict food movements to other countries in order to protect their own national cereal stocks, so food may not move from areas of higher production to those facing famine. In situations of internal conflict, restricting food supply into contested areas may be part of a war strategy.

Q4. Is the market competitive?

By identifying the number of actors at each position in the supply or value chain (producers, traders, middlemen, retailers, importers), it is possible to identify features that might distort the market. Where there are large numbers of suppliers in relation to the number of buyers, there is a competitive market, and the buyers are likely to be in a very powerful position. Distributing cash can be an effective way to meet people's basic needs. In contrast, when there are few suppliers, or traders, they can control prices and monopolise the market; in such a case, cash transfers would not be appropriate.

Q5. Is the market integrated?

Market integration allows goods to move smoothly along the supply (or value) chain from producer to consumer. It allows the demand to be met by supply. In the case of food, this would mean that the demand in food-deficit areas is met by supply from food-surplus areas. An integrated market needs good services, such as reliable flows of information, a well-developed transport system, and developed marketing networks. Without market integration, supply will not meet demand, and cash transfers will not be appropriate.

Q6. Are traders able and willing to respond to an increase in demand?

Traders' ability and willingness to respond to an increase in demand in emergency-affected areas will be influenced by the logistics and cost of supplying the affected area, and the likely reward from supplying a new market. In many instances, supplying an emergency-affected population may not be an attractive proposition for traders. Supplying unknown markets, together with limited information, can expose a trader to the risk of being undercut by other traders. The small size and short duration of markets in many cases may yield only limited rewards. Finally, populations most vulnerable to emergencies often live in remote and inaccessible areas, so transport costs are high. The factors that influence traders' response to famine are indicated in Box 3.

Box 3: Factors that influence traders' response in times of famine

Logistical constraints

Transport costs

Costs of re-directing distribution channels

Accessibility of famine-affected villages

Small surpluses available for merchants to purchase for resale

Limited rewards

Small size of famine markets

Short duration of famine markets

Opportunity cost of losing regular customers elsewhere

Limited monetary value of assets offered by peasants in exchange for food

Risk and uncertainty

Risk of being undercut by other traders

Uncertainty caused by limited information about famine markets

(Source: Devereux 1988, quoted in Harvey 2005)

Q7. What are the risks that cash will cause inflation in prices of key products?

When local markets are not able to absorb the increased demand for basic commodities, there is a risk of inflation. The answers to the questions above should indicate whether or not there is a risk of inflation. For example, if the goods that people need are available only in small quantities in the market, if there are government restrictions on the movement of goods, or traders are unwilling to respond to an increased demand, then there is likely to be a risk of inflation as a result of cash interventions.

In addition, if a cash intervention targets a high proportion of the entire population in the affected area, and/or the cash economy is relatively small, there is a risk of inflation. One way of investigating the risk of inflation posed by the amount of cash injected into the economy is to assess the volume of cash being provided by the project, compared with other inflows of cash (for example remittances).

Figure 1 on page 22 gives an example of how the questions posed in the checklist above can assist in making decisions about response to a food crisis.

Figure 1: A framework for deciding whether to distribute cash or food

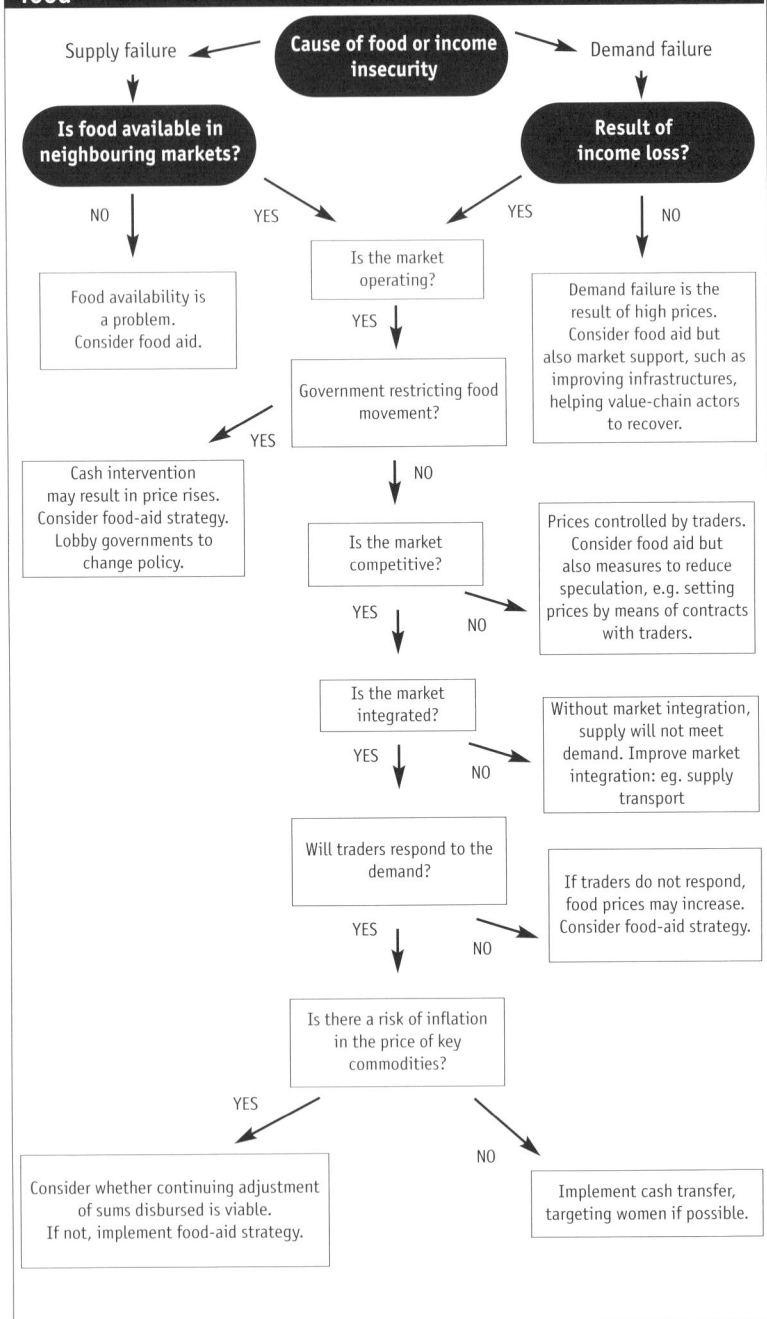

Cause of food or income insecurity

Supply failure ← → Demand failure

Is food available in neighbouring markets?

Result of income loss?

NO → Food availability is a problem. Consider food aid.

YES →

Is the market operating?

YES →

Government restricting food movement?

YES → Cash intervention may result in price rises. Consider food-aid strategy. Lobby governments to change policy.

NO →

YES → Demand failure is the result of high prices. Consider food aid but also market support, such as improving infrastructures, helping value-chain actors to recover.

NO →

Is the market competitive?

YES →

NO → Prices controlled by traders. Consider food aid but also measures to reduce speculation, e.g. setting prices by means of contracts with traders.

Is the market integrated?

YES →

NO → Without market integration, supply will not meet demand. Improve market integration: eg. supply transport

Will traders respond to the demand?

YES →

NO → If traders do not respond, food prices may increase. Consider food-aid strategy.

Is there a risk of inflation in the price of key commodities?

YES → Consider whether continuing adjustment of sums disbursed is viable. If not, implement food-aid strategy.

NO → Implement cash transfer, targeting women if possible.

A market-analysis tool

A market-analysis tool developed by ITDG (Intermediate Technology Development Group) and adapted by Oxfam may help to answer the seven questions listed above. The tool provides guidance on selecting the most important factors to investigate in order to determine whether a market functions well or not. It divides the market into three clusters. The *market environment* covers everything from infrastructure to government policies. The *value chain* considers the various parties who are involved in trading, the value that they take at each stage, and their capacity to meet the potential demand. *Market services* include transport, credit for petty traders, and information on prices and availability in key markets, all of which affect the ability of a product to reach a certain point at an affordable price. Within each of the three clusters, there is a range of variables that need to be considered to assess whether the market is functioning well. These are illustrated in Figure 2.

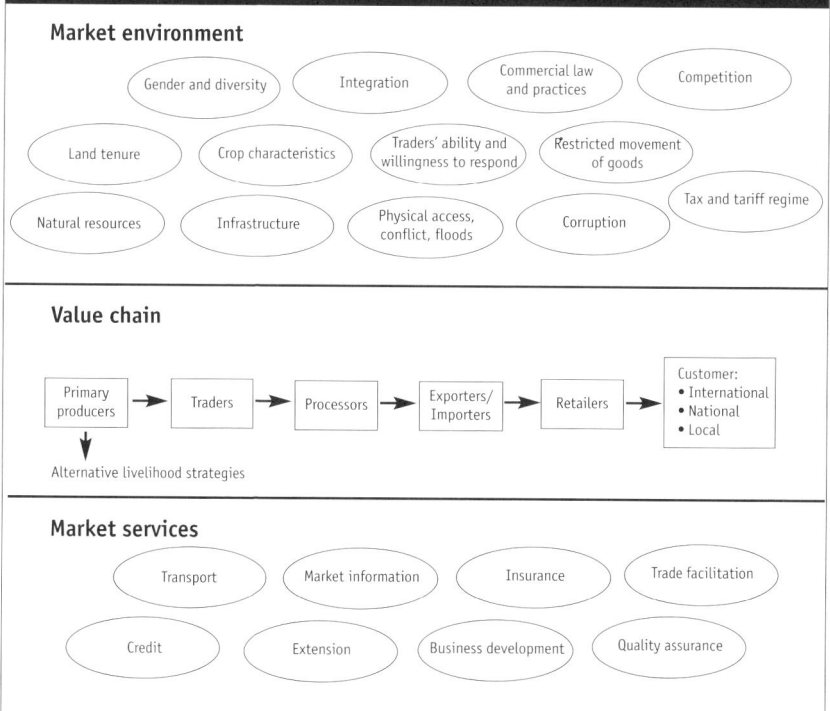

Figure 2: Market analysis tool

Market environment

- Gender and diversity
- Integration
- Commercial law and practices
- Competition
- Land tenure
- Crop characteristics
- Traders' ability and willingness to respond
- Restricted movement of goods
- Natural resources
- Infrastructure
- Physical access, conflict, floods
- Corruption
- Tax and tariff regime

Value chain

Primary producers → Traders → Processors → Exporters/ Importers → Retailers → Customer:
- International
- National
- Local

Alternative livelihood strategies

Market services

- Transport
- Market information
- Insurance
- Trade facilitation
- Credit
- Extension
- Business development
- Quality assurance

The market analysis tool can also be used to assess the impact of a disaster on the market as a whole, and therefore help to identify how humanitarian agencies can intervene to recover a functioning market. Oxfam has used the tool for this purpose only once in an emergency context, to assess the impact of floods in Haiti; see Appendix 5 for details.

Minimising risks in the delivery and distribution of cash

The assessment may identify risks of corruption, or risks to security while delivering the cash. This does not necessarily mean that you cannot implement a cash programme, because some risks can be minimised by the design and implementation of the intervention.

It is useful to consider the risks at each stage of the implementation of the cash project; this makes it easier to find ways of minimising them. (Methods of payments are considered in greater detail in Chapter 3.) Some examples of risk and ways of minimising them are given in Table 4.

Many of the same risks apply to in-kind commodity distributions, in particular food distribution. The possible risks associated with cash interventions should be compared with the risks posed by in-kind distributions. In some cases, cash may pose fewer security risks than in-kind distributions, in particular where local banks or money systems can be used. Food convoys are very visible and easy to attack, but perhaps not to loot, whereas cash deliveries are less easy to see, but more easily stolen. Any possible risks should be balanced against the benefits of providing cash, such as responding quickly, allowing choice, and stimulating the economy. In situations where leaders are not accountable, or where there is no rule of law, cash distributions should not be implemented. In many situations, however, the measures described in Table 4 should minimise the risks to acceptable levels.

Table 4: Examples of risks and ways of minimising them

Project stage	Possible risks	Ways of minimising the risks
Cash delivery	Theft, looting	Use local banks, or money-transfer companies. Make them responsible for covering losses by taking out insurance.
		Appoint monitors from the target community.
		Limit the number of people who have information about payments.
		Decentralise distribution so that smaller amounts of money are transported.
		Vary payment days.
Cash targeting and distribution	Diversion by local elites, authorities, warring parties.	Deposit money in bank accounts so beneficiaries can choose when to collect it.
	Theft	Register clearly identifiable households, or households that have been openly identified and agreed by the whole community.
		Inform the community clearly that the programme will be withdrawn if diversion of cash or threats to security occur.
		Monitor cash receipts.
Cash retention by beneficiary	Attack on way home after distribution. Taxation. Theft.	Use banks so that beneficiaries can choose when to collect money, how, and in what amounts.
		Vary payment days.
		Decentralise distribution, so that beneficiaries have a shorter distance to walk home.
		Closely monitor traditionally marginalised groups.
		Monitor the use of cash.

Deciding on the type of cash-transfer programme

The following section will help humanitarian agencies to decide which type of cash transfer, or which combination of cash transfers, is most appropriate to a particular context. The type of cash programme will vary according to the nature of the problem, the objective of the intervention, and therefore the specific target groups. Objectives might include 'to be able to meet their minimum food needs'; or 'to help farmers to re-establish their crop production'; or 'to re-establish business activities'. The programme objectives will also determine the extent to which NGOs wish to restrict the use of cash. The most common method of restricting the use of the cash transfer is through the use of vouchers whose exchange is limited to specific goods or services. In deciding on the objective and type of cash intervention, it is also important to consider the assistance provided by other agencies. For example, if food aid or non-food items are already being distributed, then the assessment needs to determine whether there are additional uncovered needs, and whether these can be met by cash transfers.

The three most common cash interventions in emergencies are cash grants, vouchers, and cash for work. Some of the factors that will influence the decision on which one to implement are discussed below.

Cash grants

A cash grant is the distribution of free cash as a relief item to targeted beneficiaries. The most common objectives of cash grants are to meet immediate food or non-food needs, or to recover productive assets. Other possible objectives include helping vulnerable households to pay off their debts or assisting in the re-establishment of businesses. Oxfam has most commonly provided cash grants to recover assets, or as part of a cash-for-work programme for poor or food-insecure households that cannot provide labour. The use of cash relief to meet basic needs should be considered more widely as part of the emergency programmes of an agency such as Oxfam, particularly in the early stages of an emergency. When cash grants are requested to re-establish businesses, one condition might be that the applicant provides a business plan. Similarly, when cash grants are given for asset recovery, the agency may ask for receipts to show that the money has been used for its intended purpose. Providing conditional cash grants is a slow process, however, and it is generally not recommended in the acute phase of an emergency, when large numbers of people are in need of assistance quickly.

Vouchers

There are two main types of voucher. One is a cash voucher, which is a voucher with a fixed cash value, and the other is a commodity voucher, which is a voucher that can be traded against a specific commodity, such as fodder or rice at a specified weight. Vouchers might be exchanged to purchase commodities from certified traders, either at distribution outlets, or in markets, or in special relief shops. The traders then reclaim the vouchers at a bank or directly from the implementing agency. Commodity vouchers protect recipients against inflation, by setting the weight of the commodity and accepting that the implementing agency will cover the cost of any inflation. Vouchers are recommended when there is an identified need for specified commodities for which local supply is appropriate and available, most commonly seeds or livestock. Voucher programmes can be used to encourage traders to enter the affected area, by providing them with a guaranteed market. In some circumstances, food vouchers may be appropriate: for example where food has been identified as the main need of the affected population. Another possible use of vouchers is for obtaining essential services: for example, the milling of relief food (as recommended to Oxfam by Creti in Darfur in 2005). When the risks of insecurity associated with transporting or distributing cash are high, voucher programmes may be appropriate because they involve fewer transactions.

Cash for work

Cash for work (CFW) is the distribution of cash in payment for work on public employment schemes, or in some cases on individual work schemes. Oxfam has applied this approach predominantly in communities where waged labour opportunities have been lost or were not prevalent. CFW interventions are commonly implemented in chronic or slow-onset emergencies, or in the rehabilitation phase of a quick-onset disaster. CFW programmes will also have objectives related to the work in hand: for example, improving the public health environment, or the rehabilitation of farms and public buildings. The projects must be necessary and appropriate if cash for work is to be appropriate. In some acute emergencies, cash for work may start as casual labour in order to get the work done quickly; in such a case, targeting the poorest or most food-insecure households starts in the second phase of the programme. Cash for work should not interfere with labour markets or with other household priorities.

The advantages and disadvantages of different types of cash intervention are given in Table 5 on page 28, which should further facilitate decision making.

Table 5: Advantages and disadvantages of different types of cash intervention

Cash for work	Vouchers	Cash grants
Advantages		
Easier to target than vouchers or cash grants	Can be directed towards food purchase and consumption	Quick to distribute and circulate
Creates community assets	Voucher exchange is easy to monitor	Minimal involvement of implementing agency at point of trade
Registering labourers for cash for work is easier than registering beneficiaries for cash grants	Less vulnerable to inflation and devaluation	Low administration costs
	Security risks are sometimes lower than for cash for work or cash grants	
Disadvantages		
High administration costs	High administration costs	Difficult to monitor usage
Some of the poor or food-insecure households may not be able to participate (e.g. elderly, ill, labour-poor households, women with other household duties)	Risk of forgery	Targeting and registration are difficult, because cash is of value to everyone
	May create a parallel economy	
	May need regular adjustment by agency to protect from inflation	
Takes up to six weeks to organise	Can take six weeks or more to organise	
May interfere with labour markets or other household activities or priorities		

While the criteria listed in Table 5 may be useful for initial decision making, it is important to realise that various kinds of cash intervention can be implemented at the same time. For example, vouchers may be used in combination with cash grants, or people may be given vouchers for work, rather than cash for work. Box 4 gives some examples of combining different types of cash intervention. In some circumstances, a combination of in-kind and cash distributions may be the most appropriate intervention. For example, in some emergencies food may not be locally available, but essential non-food items are obtainable. In this case it would be appropriate to distribute cash as a complement to food aid.

Box 4: Combining different types of cash intervention

In **Sri Lanka**, following the 2004 tsunami, the integrated public-health programme included cash-for-work projects to rehabilitate land and shelters, combined with the rehabilitation of fishing-related assets; cash grants for vulnerable households (mainly female-headed households) who had lost access to their main source of income; and vouchers to meet immediate needs for food and non-food items, once the food aid had ended.

In **Haiti** in 2004, Oxfam responded to floods and political instability, combining cash and food for work on public-works programmes with food and cash grants for vulnerable groups unable to work, through a system of shops and vouchers. The public-works schemes improved local environmental conditions by draining and cleaning canals. In **Somaliland** in 2004/05, Oxfam implemented cash for work in response to drought and chronic food insecurity, and gave cash grants to poor and food-insecure sections of the population who were unable to work.

In **Aceh** after the tsunami, cash for work was paid initially in order to dispose of solid waste, clear roads, and bury dead bodies; later it was used for farm clearing. This was more casual labour than classic CFW, because projects were not necessarily identified by the communities, and participants were selected on the basis of their motivation to work. The main aim initially was to get the work done, as well as to stimulate markets and improve people's access to food and non-food items. In addition, Oxfam gave cash grants for people to re-establish their businesses, once they had returned home.

Oxfam is increasingly using cash programmes to assist people to obtain non-food items: for example, through 'shelter shops' in Aceh in 2005 after the tsunami. Oxfam helped communities to establish local shops selling shelter materials and carpentry services. Some of the stock was salvaged from the debris left behind by the tsunami. Local people brought materials to the shop and were given cash in return. Oxfam also worked with local traders to supply other materials to the shop. Community members were given a credit limit and could acquire materials and services from the shop up to that value.

There are several innovative ways of cash programming which vary according to needs and context. As long as the basic principles of cash programming are applied, there are many different ways in which they can be implemented. The descriptions of cash grants, cash for work, and voucher programmes will give you an idea of what is possible, based on our experience; but new ways of cash programming are found on a regular basis, and we encourage you to find the most appropriate and effective ways of meeting people's needs in your particular context.

Part 2 | Implementing cash-transfer programmes

3 | Giving cash grants

Introduction

Cash grants provide project recipients with grants in the form of cash or cheques. Cash-grant interventions can target single households, or groups of households, or whole communities. Cash is usually given with one of two objectives in mind: to meet basic needs, or to help populations to re-establish their livelihoods.

Cash grants allow people to make their own choices about how to spend the money. The larger the cash grant, the more likely it is that cash will be spent on livelihood recovery in addition to meeting day-to-day basic needs. In some cases, cash grants may be given for the specific purpose of purchasing essential livelihood assets or providing initial capital for setting up a business. In all cases, expenditure should be closely monitored, to assess whether cash is actually spent in accordance with the programme objectives, and to check the accuracy of the initial assessment of needs.

Although cash grants are the most empowering type of intervention for communities, and although they maintain the dignity of beneficiaries by providing choice, such programmes are rarely implemented, because agencies and donors are cautious and suspicious of them. Fears about cash-grant interventions are the same as those listed for all cash interventions in Table 2 on page 12, but they are more extreme. Giving cash grants requires a change in the mind-set of aid workers, because it means giving control and responsibility for identifying and meeting needs to the beneficiary communities themselves. With cash grants there is an added fear of creating dependency on free handouts. If acute needs are most quickly and appropriately met by cash grants, then cash should be provided to meet these acute needs, and exit criteria should be clearly agreed with the affected community and local authorities. Long-term cash programmes are needed in most societies as part of social-welfare programmes for people living in poverty and chronic food-insecurity. Governments should take responsibility for such programmes, but in many developing countries they need support and encouragement from international agencies.

Existing information shows that cash-transfer interventions can be an effective means of alleviating emergencies. Cash grants are more efficient than food-for-work (FFW) and cash-for-work (CFW), because they can be implemented on a larger scale and more quickly, and the impacts are immediately felt.[1] The sections below are based on the experience of Oxfam GB and Novib in Somalia (2004) and Somaliland (2004/05), and Oxfam GB's experience in India (2005), Haiti (2004), Sri Lanka (2005), and Indonesia (2005).

Planning a cash-grant intervention

Plans for any cash-grant intervention should include the following steps.

- Consult other NGOs, development actors, government officials, and local leaders about the proposed programme.
- Explain the purpose of the project to the community.
- Strengthen community-based groups, or establish a relief committee.
- Recruit and train project staff – field monitors, accountants, and food security/emergency livelihoods staff – to assess and supervise and monitor the project activities.
- Develop targeting criteria.
- Set the value of the cash grant.
- Develop a system for paying the beneficiaries.
- Collect baseline information to plan and monitor the receipt, use, and impact of the grant.
- Develop a monitoring system.

Community sensitisation and organisation

In many places, cash grants are a relatively new idea. For this reason, they may be met with resistance and fear on the part of some stakeholders. This means that extensive sensitisation is needed among agency staff, local authorities, and communities to communicate the purpose and advantages of giving cash grants. Even then, managers and other relief staff may still feel uncomfortable about handing over money to beneficiaries to spend as they choose. Communities, implementing partners, and local authorities need to understand and agree on the objectives of the programme and the criteria for targeting.

Oxfam promotes the establishment of community-based relief committees. Important considerations when setting up committees and defining the role of the committees are shown in Box 5. Community relief committees are often set up or strengthened to lead the targeting, distribution, and monitoring processes. In some cases it may not be necessary to establish new committees; if they already exist – if, for example, a local committee has been established by an agency's health-promotion team – it would be more efficient to work with them. Or local government or local leaders may be sufficiently accountable and efficient that they can be entrusted to implement cash-grant programmes. Committees must represent the affected populations, so they must be gender-balanced and represent all livelihood groups, and socio-economic, religious, and ethnic groups. When supporting the communities to establish the criteria for selection, it is advisable to keep records of what criteria were used and why.

Box 5: Community-based relief committees

Key aspects of setting up relief committees:

- Committee members should be elected by the entire community. Elections should be carefully planned and monitored, to ensure that all can participate, and that those elected are people in whom the beneficiaries have confidence.
- Ensure that government officials, chiefs, and elders are well informed about the process. Invite their views.
- Ensure that clear criteria for membership of the committee are agreed in a public meeting with the community before the elections take place.
- Encourage communities to adopt criteria that include the 50:50 representation of women and men.
- Clarify the different roles and responsibilities of everyone involved, and agree together a written record of the different tasks required, with the names of those responsible.
- Encourage the distribution committee to agree its own guidelines to govern its actions in a range of scenarios, such as disruptions at the distribution point or the theft of food.

Generally, committees will have the following responsibilities:

- Disseminating information on objectives and the size of cash grants.
- Defining selection criteria.
- Selecting beneficiaries.
- Maintaining order on payment days.
- Calling names from the register.
- Receiving complaints about the programme.
- Keeping the agency informed about the operation of the system and any changes in circumstances which would make it necessary to alter the size of the cash grant.

Ensure that the committee members are aware of their accountability to the community, and ensure that the community is aware that committee members can be re-elected (or un-elected) at any time.

Implementing cash grants to meet basic needs

Criteria for targeting

Cash grants to meet basic needs should be targeted at people who are unable to meet their basic needs with their remaining food and income sources after the emergency. Grants are usually provided on a recurring basis (monthly, for example). Targets include those worst affected by the emergency, for example through destruction of their assets and savings, or loss of employment. These groups are often the poorest, or those without secure supplies of food. Cash grants can be targeted at entire communities, at households within communities, or at specific population groups, such as displaced populations and their hosts. In Somaliland, cash grants were targeted at those who risked destitution as a result of drought. They were selected by excluding households who owned more than 60 goats and/or water tanks, and those who had social capital such as access to remittances or support from their extended family.[2]

Cash grants can also be given to certain target groups as part of a cash-for-work programme. In this case, cash grants should be targeted at those unable to work, such as labour-poor households, pregnant and breast-feeding mothers, and disabled, sick, and elderly people.

Deciding the size of a cash grant to meet basic needs

To decide the size of the cash grant to meet basic needs, you should calculate the difference between the cost of the essential goods and services that households need and the cost of what they are able to acquire (or consume) from their existing sources of food and income.

Figure 3 on page 36 shows a hypothetical household's needs for essential food and non-food items, compared with what they can obtain from their existing sources of food and income. Setting the level of cash grants does not necessarily require such a precise quantification as indicated in the figure, but it is advisable at least to consider the extent to which households are able to meet their needs through their own means. If a food-security assessment has been conducted, it should provide information on changes in people's sources of food and income; but such changes are usually not quantified. Calculating a cash grant, however, does require quantification.

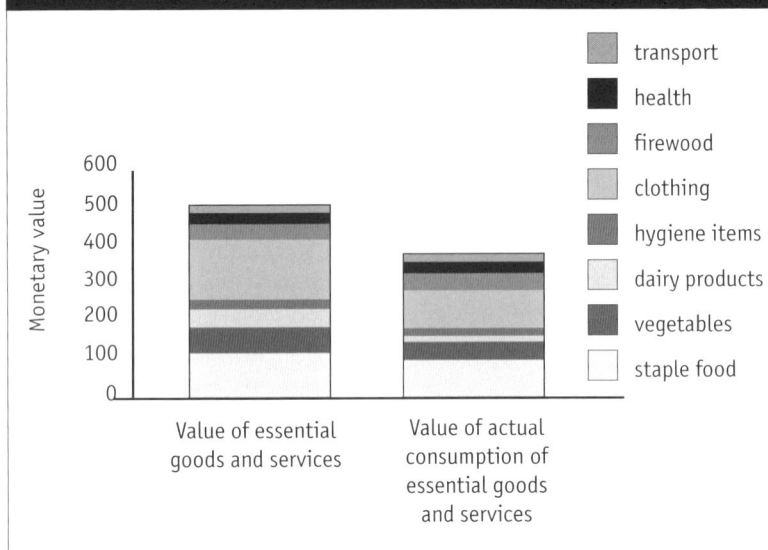

Figure 3: A sample household's needs for food and non-food items

Legend:
- transport
- health
- firewood
- clothing
- hygiene items
- dairy products
- vegetables
- staple food

Y-axis: Monetary value (0, 100, 200, 300, 400, 500, 600)

X-axis categories:
- Value of essential goods and services
- Value of actual consumption of essential goods and services

Here are some other issues to consider when setting the level of a cash grant:

- Goods and services essential for meeting immediate needs include staple foods (cereals, pulses or meat/fish, oil), vegetables or other foods rich in micronutrients, hygiene items, shelter, kitchen utensils, health care, water, firewood, clothing, and transport.

- Analysing people's expenditure can give a good indication of what they consider to be their priorities, and therefore the most essential goods and services.

- Consider how prices are likely to change during the programme. The programme itself may lead to some inflation, as a result of the increased demand generated by the cash transfers.

- When assessing household consumption of essential goods and services, you should include any assistance received through other humanitarian and social protection programmes.

- Households may be using coping strategies which are damaging to their livelihoods, health, and dignity, in an attempt to meet their needs. Such strategies should not be considered as part of this analysis.

- The provision of cash to meet a specific need, such as food, may not have the desired impact if the household is not meeting its other essential needs: in other words, the cash may be used for a purpose for which it was not intended.

In Somaliland in 2004/05, the cash grant provided to the poorest households was $50 per month, and was based on the cost of food, sugar, oil, water for livestock and humans, drugs for humans and livestock, and relocation costs for livestock. These had been calculated as the basic needs of the poorest or destitute populations. Since the target group were destitute, i.e. unable to meet any of their needs through their own means, the value of the cash grant was equal to the total value of the essential goods and services.

Implementing cash grants to rehabilitate livelihoods

Criteria for targeting

An emergency assessment or food-security assessment should identify the effects of the emergency on various different livelihood groups, and their needs and priorities. Different types and sizes of grant may be targeted at different livelihood groups, depending on the nature and extent of their assets, savings, and income lost as a result of the emergency. Further assessments will be needed to identify the types and quantities of asset that the targeted population require in order to resume their normal livelihoods. Grants for livelihood recovery are likely to be given on a one-off basis.

Oxfam provided grants to re-establish livelihoods in India, Sri Lanka, and Indonesia as part of its response to the tsunami that struck in December 2004. The main aim of the cash grants was to enable livelihoods groups to purchase the assets needed to re-start their income-earning activities. In India, Oxfam provided widows and households headed by women with cash grants to re-establish food shops, tea shops, and vegetable gardens, and to buy fishing nets. The women needed some starting capital, and bicycles to enable them to deliver products to local markets.

In Sri Lanka, cash grants were used to help to re-establish people's previous businesses, or to buy productive assets such as livestock. The three main target groups were people engaged in small businesses, farmers, and people engaged in lace making and the coir industry (making items such as baskets out of rope).

Deciding the size of a cash grant to rehabilitate livelihoods

The size of a cash grant which aims to rehabilitate livelihoods depends on whether basic needs as well as livelihood-recovery needs have to be met, and the cost of the assets and materials that need replacing in order to re-establish former livelihoods. If basic needs are not met by other interventions, then it is likely that a substantial proportion of any cash grant will be used to cover basic needs. The cash grant should cover both the basic needs and the costs of rehabilitating livelihoods. As shown in Chapter 1, the larger the cash grant, the more likely it is that beneficiaries will spend it to recover their livelihoods, prioritising such things as the purchase of small stock. The cost of basic needs should be worked out as indicated in the previous section, and the cost of livelihood assets or materials will need to be added to this calculation.

In tsunami-affected areas of Sri Lanka and Indonesia, Oxfam provided cash grants specifically to recover livelihoods, because basic needs were already met through other interventions. So the size of the cash grant was determined by the cost of the assets or materials required. For example, lace makers received a grant of 5000 Rupees[3] per household to purchase lace-making equipment and the cotton needed to make the lace; coir workers received a grant of 8000 Rupees between three women to buy a coir-making machine (operated by three women), and a preliminary stock of raw coir. Farmers received a grant of 20,000–30,000 Rupees to replace livestock lost during the tsunami. In Indonesia, the cash grant aimed to cover the cost of shelter, workshops, tools, equipment, and raw materials for small businesses. The implementation process for providing cash grants for small businesses is described below.

Implementation process

The following procedure should be followed for making cash grants to buy materials to re-establish small businesses:

- Invite proposals from individuals and groups who need support to restart their business activities.
- Limit proposals or business plans to one per household.
- Ensure that men and women have equal opportunities to apply for grants and loans.
- Ask applicants to demonstrate that they have experience, or at least recognised potential for success, in the desired business.
- Decide on the amount of cash that can be granted; it usually varies according to whether the application comes from an individual or a group, and according to the type of business. The grants should be limited to the purchase of essential items needed to re-start businesses.

- Beneficiaries apply for grants, presenting business plans which should include the types of activity that they plan to start or re-establish, the type of equipment and raw material needed to attain an identified level of production, the labour available, further skills required, and the approximate costs of all inputs.
- Verify the costs by assessing current prices in markets and shops.
- Reach agreement on the total amount of the grant.
- Sign a formal contract with the implementing agency.
- Ask the recipients to provide receipts as proof of purchase against the items listed in the signed agreement.
- Monitor the use of the cash.

Beneficiaries should already have the skills to run the business and should therefore not need basic training. When possible, such cash grants should be accompanied by training in business management, improving the profitability of the business, and developing new ways of generating income. Beneficiaries might also be supported to gain better access to markets and traders.

Methods of cash delivery and payment

Identifying appropriate and safe methods to deliver and pay the cash is an important step in planning cash programmes. The aim is to reduce risks to the relief agency, when transporting and distributing cash; to reduce risks to the project recipients after the distribution; and to reduce the management load on the implementing agency. There are three options for transferring the cash: the local banking system, money-transfer companies or institutions, and direct delivery and payment by the implementing agency. The three options are discussed in turn below. An initial assessment (see Table 3 on page 16) should therefore determine whether banks are functioning and accessible to the affected population, and whether there are any other financial transfer mechanisms that can be used to deliver cash safely.

Using local banking systems

When a banking system is present, and accessible to the project recipients, relief agencies can either make payments into bank accounts or contract bank officials to make the payments. Paying into bank accounts has the advantage of being safe, of introducing recipients to formal bank systems, and giving them the means of withdrawing money when it is convenient to them. Bank accounts can also be a way to promote saving, since recipients can save money in their account to cover expected seasonal expenses. Bank accounts are safer for the recipients, who thereby do not have to keep cash at home, and also for project staff, who do not have to handle cash directly. The bank account system considerably reduces staff workload and ensures documentation and proof of payment. Banks can be contracted to provide mobile services, thus reducing the risk of corruption and leakage, as banks are usually considered trustworthy, and banks will have their own 'cash in transit' insurance. The disadvantages are that banks usually require some days to prepare the disbursements, and cannot always be flexible in the timing of the distribution. Mobile banking systems are problematic for people who are not available on pay day, if a bank official cannot come at any other time.

Box 6: Using local banks to disburse cash

After the earthquake in **Bam** (**Iran**), the government set up bank accounts for beneficiaries, to which cash was transferred directly.

In **Zambia**, GTZ opened bank accounts for recipients living near the local town, while for those living more than 15km from the town payment points were set up in schools and health centres.[4]

In **Ingushetia**, SDC made use of the postal bank system to transfer money to the project recipients.[5]

In **Indonesia** (Aceh), Save the Children negotiated a contract for cash distribution with a local bank. The bank was responsible for cash management, and bank cashiers made payments every week or fortnight. Project staff monitored the timeliness of payments, the disbursement of the cash, and the use of the cash. Save the Children paid the bank for the service, including 2 per cent for insurance, paid the salary of the bank cashier, and provided vehicles and drivers. At the same time, Save the Children distributed cash directly in those areas that were not accessible by the bank.[6]

In **Mozambique**, an NGO contracted one of the country's banks to provide mobile banking, and a local security firm to enhance security at distribution sites.[7]

In **Jamaica** in 2004, Oxfam provided beneficiaries with nominal cheques which could be redeemed in local banks.

Using local money-transfer companies

In contexts where there are no formal banking systems, some relief agencies have developed innovative ways to distribute cash. These methods are based on local traditional systems and therefore require a good knowledge of the local context.

In Somalia and Somaliland, agencies have used the local money-transfer system, usually used for remittances, to distribute cash.[8] These companies took a 5 per cent fee and they accepted responsibility for any loss. In Afghanistan, Mercy Corps devised a method which uses the local 'banking' system (*Hawala*) to transfer the relatively large sums required to meet payroll needs in the field.[9] Paymasters transferred the payroll cash to group leaders, who paid individual labourers, with Mercy Corps project engineers providing oversight. The *Hawala* system ensures that transported money is the responsibility of the money changer, while also boosting traditional systems of cash transfers.[10]

In Haiti, Oxfam GB made use of local shops to pay cash grants as well as cash-for-work wages on a fortnightly basis. The use of shops prevented a recurrence of security problems that had been previously experienced during direct distributions of commodities.

Direct payments by implementing agency

If using local banks or money-transfer companies is not feasible, or does not appear to be the most appropriate option, then it may be necessary to plan and make the payments directly. Several aspects of making payments have to be planned in advance. They include the following considerations.

- Money often needs to be ordered from banks in advance. Order small denominations, and try to get the same denominations for each beneficiary, for ease of counting and distribution.

- Counting money takes time. Individual cash payments should be counted and packed before the day of payment. Money, once counted, should be kept in a sealed box in a safe at the bank until payment day.

- To speed up the process, attendance sheets and other documentation should be collected in the afternoon or evening before the planned payment, so that programme staff can verify the records before any pay-out.

- Payment sheets need to be produced for each community, with a name and identity number if possible. A separate sheet should be completed for each community, for ease of checking the number of payments made.

For making payments, lists of beneficiaries should be agreed with local community committees and checked on the basis of the agreed criteria. Beneficiaries should receive an identification card; if this is not possible, ensure that a community member or staff member from that village can identify each beneficiary. If the cash grants are delivered in the community, you should select a safe, controlled location, and ensure that distributions are made to small groups of recipients. Direct cash payments are usually conducted in the following manner.

- Beneficiaries are called three at a time, to avoid accusations of favouritism or other wrongdoing.
- Literate beneficiaries are selected to count money and make sure that beneficiaries understand the denomination received.
- Beneficiaries must sign their names or make a fingerprint to confirm receipt of the money.
- All the payment sheets should be countersigned by field officers, partner staff, and local authorities.

Staff monitors, together with relief committees where appropriate, are responsible for supervising the identification and verification of beneficiaries during distributions; for mediating and resolving conflicts among community members; and for facilitating co-ordination with the community. At the end of the disbursement, a witness from the community should sign the payment sheet to verify that the payment was made.

The steps that should be taken to minimise the security risks of direct cash payments are summarised in Box 7.

Box 7: Minimising the security risks associated with direct cash delivery and payments

- Consider taking out insurance cover against the risk of loss while transporting cash to projects in areas where there are no banks.

- All stakeholders in the community (including elders, politicians, and non-recipients) should be informed about how payments will be made.

- Ensure that the community understands the consequences of any threat to security: in other words, that programmes will be withdrawn or suspended if necessary. Communities will protect you in order to protect themselves.

- Limit the number of people who have information about payments. Only two or three people in the agency should have access to information about the date and time when a payment is to be made. Beneficiaries in the field should not know in advance when a payment is to be made. Long-standing or local staff should be involved in making payments: this reduces the risk of theft.

- Decentralise distribution as far as possible, so that smaller amounts of money are transported to different locations, and beneficiaries have a shorter distance to walk home.

- Make payments on a random basis. Do not always set off to make payments on the day after withdrawing the money from the bank. Vary the locations of payments if possible, especially in towns. Vary the routes of staff carrying money to and from the field. Vary the individuals making payments on different project sites.

- When transferring cash by car, divide the money into two or more bundles and hide them in different parts of the car: attackers may leave once some money has been surrendered to them. Ensure that the vehicle has a high-frequency radio for communication.

- Ensure that payments are completed in time for beneficiaries to reach their homes during daylight.

- Avoid spending the night at the project site when disbursing cash, even if you have finished making the payments. Travelling on the following day might encourage the assumption that you have more cash.

Monitoring and evaluation

This section provides information on indicators to monitor the process and the impact of a cash-transfer programme. Further indicators specific to monitoring cash-for-work and vouchers interventions will be discussed in Chapters 4 and 5. Samples of monitoring checklists and questionnaires are provided in the appendices.

Monitoring and evaluation are vital for programme accountability and learning purposes. Properly done, they will ensure that the programme is appropriate and relevant to people's needs; that it is implemented in the way that was intended (*process monitoring*) – in other words, that it is efficient and effective; and that it is having the intended impact and is minimising negative impacts more effectively and efficiently than other types of programme (*impact monitoring/evaluation*).

Baseline information

Baseline information is needed at the beginning of the project in order to monitor the changes that it brings about; such information includes details about incomes, expenditures, and assets. Baseline data may be already available, either from secondary sources, such as early warning and food-security monitoring systems, or from emergency assessments. When such data are not available, a baseline survey should be carried out on a sample of the targeted population.

A baseline survey should provide information about the previous state of affairs (usually referred to as 'normal years') and the situation at the beginning of the project. To obtain information about a normal year, target groups are asked to recall their living conditions before the project began. Methods to collect baseline data are similar to those used in food-security or household-economy assessments.

Baseline surveys are usually conducted on a representative sample. The accuracy of the sample will depend on the size and how we select it. The cost and time available are usually the two factors that determine the sample size. A baseline survey conducted in an emergency context does not need to be a statistically perfect study. However, the sample must be large enough to inspire confidence that it is fairly representative of the majority of the population, without wasting too many resources and time in an emergency.

If possible, include people who are not direct beneficiaries of the project, in order to form a broader picture of the impact of the cash intervention on the entire population. The information collected in a baseline survey is an important resource for the setting of process and impact indicators.

Below is a summary of the information that is usually collected in baseline surveys for cash-transfer interventions. It is only general and indicative: specific information will probably be needed, case-by-case.

Household information

- What are people's sources of food and income at the start of the project, and in a normal year?
- What are the average income and expenditure of different groups within the population at the start of the project, and in a normal year?
- What are the key assets of various livelihood groups at the start of the project, and in a normal year?
- What do people commonly buy and sell at the start of the project, and in a normal year?
- Do they normally obtain food through purchase?
- What coping strategies do people usually adopt in periods of food scarcity?
- What are the normal migration patterns?
- Do members of the household migrate for work at certain times of the year?
- What is the general level of people's knowledge of business?
- What are the characteristics of gender and social/ethnic relations at household and community levels?
- What would people do if they had more money?
- What are the normal intra-household mechanisms concerning the management of cash and decisions on expenditure? Who keeps the money in the household, and who decides how to spend it?
- What is the level of household debt? What is people's usual access to credit and banks?
- Who is responsible for debt repayments?

Market information

- What food is available on the market, and in what seasons?
- What are the prices of essential food and non-food items, and what are their seasonal variations?
- What is the level of trading? Who are the actors in the market supply chain? How large are the numbers of suppliers, middlemen, and final consumers? What is their power?
- Who controls trade? Is the market controlled by one or a few big traders?

- What market services are usually available? Is there access to credit? Is information on market prices and availability accessible? Are insurance mechanisms in place?

Monitoring

Some indicators should be monitored on a regular basis throughout the duration of the programme, to see whether it remains relevant, whether it is being implemented as intended, and whether it is having the expected impact. These indicators should be stated in logical frameworks. Regular monitoring is necessary to make sure that the programme is adapted or changed if it is not relevant, that it is implemented according to plan, and that it is having an impact. This is especially important if a programme provides regular cash payments for a period longer than three months. Mechanisms should be established to make sure that regular information is collected, analysed, and acted upon: for example, by means of regular reporting and meetings between stakeholders (programme staff, partner staff, community committees, beneficiaries, non-beneficiaries, suppliers of goods and services, etc.). The minimum set of indicators to monitor are as follows.

Process indicators

- Did the beneficiaries / suppliers receive the correct sums of money?
- Was the payment made on time?
- Were beneficiaries and other stakeholders satisfied with the process and methods of implementation?
- What other relief assistance are cash beneficiaries receiving?

Impact/outcome indicators

- How much have income and expenditure changed since the start of the cash programme?
- How have sources of food and income, and coping strategies, changed?
- What was the additional income used for? What did people purchase?
- Were items that households wanted to buy available in the market?
- What changes took place in market prices of key commodities?

Evaluation

The following information should be collected in evaluations, and the regular monitoring information should form the basis of any evaluation.

Appropriateness

- How were the needs of the population assessed? Was it a food-security or emergency assessment? Did it include an identification of the most vulnerable and/or worst-affected groups and the most appropriate interventions for each group?
- Was a market analysis conducted? Was the project design based on a good understanding of demand and supply for basic goods?
- Were community representatives and key stakeholders involved in the needs analysis and design of the programme?
- What were the needs of the population, and was a cash intervention the most appropriate means of meeting those needs?
- Was the intervention justified on the basis of an analysis of need?
- Were community perceptions of cash programming, and past experience with cash programming, taken into account? Does the community think that cash was the best response?
- Were the criteria for targeting beneficiaries appropriate, and did they relate to the assessment findings and the objectives of providing cash grants?
- How was the value of the cash grant determined? Did the process take account of people's existing income, coping strategies, and household debt? How did the cash value relate to the objectives of the programme?
- Was a risks analysis carried out before starting the project?

Coverage

- Did the project cover the worst-affected areas/populations? Were any groups of people left without assistance?
- What proportion of the affected population/area was targeted?
- What proportion of the target population received cash?
- To what extent did the programme meet the needs of the most vulnerable in the population?
- Was targeting carried out as planned? Were there any errors of inclusion or exclusion? (In other words, were any people included in the programme who should not have been included, or were any people excluded who should have been included?)
- What proportion of the beneficiary population were women?
- What is the community's perception of the coverage?

Connectedness

- How were local resources and capacities strengthened in order to respond more effectively in the future?
- How did the project take account of existing capacity, both of your agency and local institutions (government and civil society)?
- Did the project take into account existing social safety-net mechanisms?
- How were the cash interventions linked with other livelihood-support interventions, including other short-term emergency responses and longer-term livelihood support?

Coherence and co-ordination

- How were the cash interventions co-ordinated with the programmes of other organisations or government agencies working on similar projects or in the same area?
- What was the level of co-ordination between agencies? How did the project take account of assistance being provided by other agencies?
- If the intervention was carried out through partners, how well did they co-ordinate their work with that of other actors? Was the intervention suited to their capacity?
- How did the response relate to government policies and strategies?
- How actively did the community participate in the planning, implementation, and monitoring of the programme?

Efficiency

- Was there a difference between the planned costs provided for in the project budget (staff needs, materials, running costs) and the actual costs of implementing the programme?
- What were the main constraints in achieving the project within the planned budget?
- What method was used for paying the cash, and was this the most efficient and safe method?

Effectiveness/implementation

- How many new staff or partners were hired for the project, what were their roles, and how were they trained for the project?
- To what extent were community representatives and other key stakeholders involved in the implementation of the programme?
- Were relief committees established for targeting and implementation, and did the community understand and accept the role of the committee?

- Which household members received the cash, and what were the reasons for this?
- Did beneficiaries receive the cash on time?
- Did beneficiaries receive the correct amounts of cash?
- How many beneficiaries received cash, and how does this compare with the target?
- Was a monitoring system set up? Did it include indicators to monitor relevance, implementation, and impact?
- What is the community's perception of the payment process, in terms of timing, amounts, location, and method?
- How did beneficiaries use the additional cash income?
- Was the value of the cash transfer sufficient to meet the objectives of the programme?
- Was the timing of the project appropriate for meeting the identified needs?
- Were beneficiaries able to access goods and services in the required quantity and of the required quality?

Impact

Food and income security

- How did beneficiary households use the additional cash?
- What proportion of average household income was provided by the project?
- What were the changes in sources of food and income, and asset levels? (Try to compare conditions in a normal year with post-disaster and post-project conditions.)
- What were the changes in expenditure? ((Try to compare conditions in a normal year with post-disaster and post-project conditions.)
- What were the changes in debt levels and coping strategies (including migration)?
- Was there an impact on employment, labour, production systems?
- Did beneficiaries face any constraints in the way they used cash? How could these be minimised?

Markets

- Did the programme have an impact on market prices, employment patterns, and labour availability in the area?
- Did the project affect the availability of goods, both locally and at a wider level?
- Did the programme have an impact on trader activity, or control over trade in the market?

Social impact

- If women were targeted, what was the impact on gender relations in the household and the community?
- What was the impact on control of cash resources and expenditure within the household?
- Was there an impact on social relations between groups? Did any conflict arise between households/areas that were targeted and those that were not?

Security

- Did the project have an impact on security for the implementing agency or the beneficiaries?
- What measures were taken to minimise security risks?

General

- Were there any problems or negative impacts associated with the programme? (Take care to consult all key stakeholders, include community members.)
- Are the positive changes that have been achieved likely to be sustained?

Cost-efficiency

There are two options for calculating cost-efficiency:

- Either compare the cost of cash transfers and in-kind distributions. For example, calculate the cost of distributing food aid and compare it with the cost of cash transfers. This should include procurement, transport, delivery, registration, and staffing (administration), as well as the value of the food aid or cash provided.
- Or calculate the administration costs of the intervention and the proportion of funds that went directly to the beneficiary.

An estimation of cost-effectiveness would require an added analysis of which intervention was most effective in meeting the needs of the beneficiaries.

4 | How to implement cash-for-work programmes

Basic principles of cash for work

Cash for work (CFW) is the distribution of cash in payment for work that is done either on individual projects or on public works schemes. CFW has a dual objective:

• To assist in meeting basic needs and improve livelihoods by improving purchasing power.

• To provide an asset for a community or a household. Ideally this asset should improve the livelihoods or environment of the community as a whole.

Other objectives have included the intention to contribute to economic recovery by boosting local business through increased demand for goods and enhanced ability to invest in business.

Cash for work is more appropriate than food for work (FFW) when food is available and local markets are functioning. CFW is often implemented together with a range of other cash interventions (cash grants and vouchers), food aid, and food-security interventions or livelihoods interventions (to support livestock rearing, fishing, and agriculture).

Oxfam has implemented CFW programmes in a wide variety of contexts: for example as part of drought mitigation and recovery, and in response to chronic food insecurity or seasonal food insecurity. CFW has also been implemented following floods (and the Indian Ocean tsunami), in clean-up operations after hurricanes and cyclones, or for water and sanitation programmes and solid-waste disposal in IDP camps.

The basic criteria for implementing cash-for-work programmes, discussed in Chapter 2, may be summarised as follows:

• There should be no absolute shortage of basic commodities, essential food items, and fuel.

- Markets should be functioning and accessible.
- Food insecurity and income insecurity are the result of loss of employment or loss of assets.
- Rebuilding assets and/or community infrastructure, and cleaning up debris, are an essential part of the emergency operations and are needed to rebuild livelihoods.
- There is potential scope for community organisation.
- There is potential for accountable community representation.
- Projects should not interfere with labour markets or undermine other household activities.

There may be circumstances where it is not possible to be certain that these conditions are in place before starting a CFW project. In this case, it is best to start with a small pilot project. If this works well, the project can be expanded later, or in the next crisis that affects the community. This approach was adopted by Oxfam in Kenya, where the first cash-for-work programme was implemented in Wajir in 1998 for 13,000 beneficiaries; three years later Oxfam targeted 70,000 beneficiaries in both Turkana and Wajir.

Table 6 lists the principles of CFW programming. There may be some exceptional cases where it may be more important to start the project as soon as possible, rather than wait until all the criteria can be fulfilled. For example, if the project consists of latrine construction or solid-waste disposal in camps, then it is best to start the project by employing casual labour. The difference in objectives between casual labour and cash for work can be summarised as follows:

- **Casual labour:** labour is hired to get a job done quickly and effectively. For example, there may be an urgent need for latrine construction or solid-waste disposal in a camp because of health risks. The priority is to get the job done. Criteria for employment should be experience in the required job, or the ability to gain it quickly.
- **Cash for work:** households are provided with employment opportunities to give them cash to meet their basic needs and improve their livelihoods. The work is targeted at the poorest or most food-insecure households, and efforts should be made to include women.

In some situations, doing the work and providing income may be equally important, and needed quickly on a large scale. This was the case for many tsunami-affected populations. In Aceh, Indonesia, Oxfam initiated projects that were open to anyone willing to work; in some cases, whole villages were employed. This approach was successful in assisting people's quick return to their villages (by clearing roads, repairing bridges, and clearing debris from land), as well as providing an income. In some circumstances, therefore, it may be appropriate to initiate a project as casual labour and move on to cash for work at a later stage. It is important that the objectives of the programme are clear in each case, with clear criteria to identify when or if CFW programmes can be started.

Table 6: Principles of cash-for-work programming

Principle	Explanation
The most food-insecure or the poorest people should be targeted	The beneficiaries of the programme should be those who have lost a large proportion of their food or income sources as a result of the disaster. They are identified through a community-based targeting process. Setting pay levels below the minimum wage may promote self-targeting in favour of the most poor.
The most physically vulnerable people should be included (including HIV-affected households)	Arrangements are made for those unable to work: for example they could be given cash grants or vouchers instead of CFW; or they may be permitted to nominate someone to work on their behalf; or light work might be offered to them.
The community should 'own' the programme	The community identifies project activities. This involves a process of community mobilisation to raise awareness about the nature and process of CFW programming. The identification of working units in general, and work for disabled or elderly community members in particular, needs to be facilitated with the community.
Work should be labour-intensive	Programmes should employ as much unskilled labour as possible, to maximise the impact on the largest possible number of affected households. Care should be taken not to undermine normal voluntary community activities.
A gender balance should be ensured	Projects should promote female participation. Often CFW projects aim to ensure that at least 50 per cent of the beneficiaries are women. A variety of activities should be implemented, a majority of which will be suitable for both men and women. Child-care facilities may be needed.
Equal pay should be the rule	Women and men will be paid equally for agreed units of work.
Essential livelihood activities should not be undermined	CFW activities should not interfere with or replace traditional livelihoods and coping strategies, or divert household resources from other productive activities already in place.

Planning the intervention

Once the decision to initiate a cash-for-work programme has been taken, certain essential steps should be followed, as listed below. Some of the steps are the same as for cash grants, but they are repeated here for ease of reference.

1 Consult other NGOs, government officials, and local leaders about the implementation of CFW programmes. In many places, CFW is a relatively new idea and it may be met with resistance and fear on the part of some stakeholders.

2 Ensure that the communities understand the purpose of the CFW project and the need for the whole community (not only the leaders) to be involved in identifying the most appropriate projects.

3 Recruit and train project staff, who should include field monitors, a logistician, and technical staff to assess and supervise work on projects.

4 Develop targeting criteria, and ways of providing assistance to those who are unable to work. Set age limits for people participating in the work.

5 Identify community committees to help with targeting, supervision, and monitoring of the project. Ensure that the ethnic, religious, political, and gender composition of the committee represents the community.

6 Collect information to assist planning. This should include information on community organisation and representation, logistics, and the labour economy. See Box 8 on page 56 for examples of information to collect.

7 Facilitate initial decision making by the community. Ultimately, the community will decide on the most relevant and useful CFW scheme, working units, timeframes, wages, mode and interval of payments, etc.

8 Discuss the practical details with field staff and/or partner organisations.

9 Select the project site and verify technical specifications. For some projects, it will be necessary to carry out considerable design work in advance, and to ensure that the proposed project will do no harm to other communities or individuals; for example, rehabilitation of irrigation canals may have negative impacts downstream on other communities.

10 Once the projects have been verified, it is possible to begin working simultaneously on the following tasks:

- selection of beneficiaries

- identification of work norms: working days per week, hours per day, and deadlines by which to complete the work (to confirm beneficiary numbers and size of work units)

- ordering of tools, or other necessary items; for example, vouchers, beneficiary cards, etc.

- preparation of monitoring and payment formats, and their translation into local languages

- application for 'non-objection' certificates from government and other official institutions or representatives, if work on State roads or land is planned.

11 Consult technical experts for advice on how to calculate and define work units and areas of coverage.

12 Offer on-site technical guidance for monitors and relief committees responsible for project supervision. This can be done at the same time as project demarcation (for example, for a water-harvesting scheme).

13 Distribute tools, beneficiary cards, etc.

14 Collect baseline information and set up a monitoring system.

A clear timeframe needs to be determined at the point when the project proposal is formulated. Identification of projects and beneficiaries, if the stages listed above are followed, may take as long as 4–6 weeks. This is usually appropriate, because CFW is not intended as a life-saving intervention at the initial stage of an acute emergency, and is often preceded by cash grants or food aid.

Arrange meetings with all key stakeholders to explain the programme. Writing a summary of the programme and distributing it to NGOs and authorities is one way to avoid conflicting objectives and duplication of work. Appendix 6 is an example of an information sheet produced by the Oxfam GB team in Turkana, Kenya.

Box 8: Key questions for planners of cash-for-work projects

Community organisation

- How reliable is the local administration system and traditional leadership? Does the community trust them?

- What committees already exist?

- How cohesive is the community? Are people used to working together? How long have they lived together as a community? Are there any economic, cultural, religious or political differences within the community?

- What is the attitude of the community towards paid labour? Does any part of the community oppose it?

- What are the existing support systems for the socially vulnerable?

Logistics

- How good is access to the area?

- What is the current state of communications in the area?

- Are materials available locally?

- Are tools and equipment available locally?

- What storage facilities exist?

- What are the security risks for transporting and storing materials?

- What are the distances and travel times between communities?

Economic information

This information should be collected in addition to the economic information listed in Chapter 3 under 'Monitoring and evaluation'.

- What kind of work is normally done within the community? (Include agriculture and livestock-related work, as well as wage labour, civil service, and self-employed income-earning strategies.)

- Do normal work activities vary with the seasons?

- What is the agricultural calendar? When are people expected to be working on their land? When is food availability at its lowest?

- What is the current availability of employment for members of the community, and how is this likely to change over the coming months?

- What are the gender roles that apply to work within and outside the home?

- Do households normally migrate for work? Is this seasonal?

- What are casual-labour rates for community members?

- What is the government minimum wage?

Determining projects

Key aspects to consider when determining the projects include the following:

- What standard of technical design and construction is required (i.e. does it require skilled or unskilled labour)?
- Will the project create a community asset (water pans, new roads, shelter, etc.)?
- Has the community selected the project? Note that communities who have a history of receiving assistance will have pre-conceived ideas about the activities that Oxfam (or other agencies) will support. Make sure that the activities identified will address problems within the community.
- Does the project provide access to a new facility or service previously unavailable (access to clinic, market, etc.)?
- Does the community have the capacity to operate and maintain the facility or asset in the long term?
- What impact will this project have on the environment?
- What impact will the project have on the normal labour market, including labour rates?
- What impact will the project have on existing workloads and access to other productive opportunities?
- How does the timing of the project relate to the agricultural calendar, in terms of labour requirements and the likely inflationary effect of cash when food availability is low?

The projects need to be labour-intensive, in order to maximise participation and transfer the appropriate amount of cash. However, if the manual labour is very heavy, workers may expend too many calories and it may be difficult to target vulnerable groups. The project should require skills that are widely available locally. Table 7 on page 58 gives examples of the types of project that Oxfam has implemented in the past.

It is important to include a range of activities in which different groups of people can take part, to ensure the maximum possible inclusion of vulnerable people. The following factors should be taken into consideration:

- The commitment of time required for CFW, (a) in terms of daily scheduling, and (b) in terms of the overall duration of the project.
- The type of work to be done: is it light or heavy? Is it culturally acceptable, especially with regard to any division of labour traditionally determined by gender or ethnic factors? Who is unable to carry out heavy manual labour?

Table 7: Examples of Oxfam cash-for-work projects

Country	Disaster	CFW projects
Uganda (2001)	Conflict	Road construction Dam construction De-silting wells Shelter (improved housing)
Bangladesh and Pakistan (2001)	Floods	Individual and community structures Road repair and reconstruction Raising ground-levels (to protect roads, schools, markets, mosques, health centres, cluster villages, and community flood shelters) Community flood shelters De-silting ponds and water tanks Shelter reconstruction Bunding to raise houses and livestock shelters Construction of raised housing and livestock shelters
Afghanistan (2003)	Chronic food insecurity	Building water reservoirs Building walls to protect against erosion Tree planting Fodder collection and planting Embroidery

Country	Disaster	CFW projects
Cambodia (2003), Kenya (2001), Eritrea (2003)	Drought	Road reconstruction Pond digging/ clearing Bund replacement Land management – terracing Production of housing materials Construction of community centres Road clearing/ rehabilitation Construction of refuse pits Construction of night-soil disposal pits Pan or dam de-silting Improvement of shallow wells Construction of reservoirs Construction of troughs for watering animals
Grenada (2004)	Hurricane	Farm clearance (stones)
Philippines (2005)	Cyclone	Road clearing Farm rehabilitation: clearing silt from fields
Haiti (2004)	Conflict and floods	Riverbank rehabilitation Road construction Canal cleaning
Sri Lanka/ Indonesia/ Maldives / India (2005)	Tsunami	Latrine building Drainage canals Solid-waste disposal Road clearing Cleaning public buildings Village clearing (drains, roads, rubble) Agricultural land rehabilitation Brick making

continued ...

Country	Disaster	CFW projects
		Boat / engine repairs
		De-silting of agricultural land
		De-silting of ponds
		Tree plantations
		Shelter construction
		Reconstruction of salt pans
		Lagoon clearing

Selecting beneficiaries

The number of households to be included in a CFW programme depends on the extent of the need and the estimated number of people required to complete the work. Often the size and type of work are based on the number of beneficiaries for targeting. Proposals should state both the number of beneficiaries to be targeted and the amount and type of work to be done.

The number of project beneficiaries should be calculated according to the following factors:

- the geographical area affected
- the livelihood groups affected
- the number of vulnerable households within a livelihood group (in the case of internally displaced people and returnees, a programme might attempt to include at least one member of every household)
- whether or not part or all of the need is being met by any other agencies, or the government.

Lack of resources usually means that NGOs must focus on the most severely affected households, livelihood groups, and geographical area. It is unlikely that any activity can target the entire population, so some form of selection is needed. This is usually done through community-based targeting (see Appendix 7). Other methods of targeting include self-targeting, through setting pay below the minimum wage (so that only the poorest will apply for work). Oxfam recommends the setting of wage levels based on needs, and on comparison with normal wage levels. Pay levels should be set below the minimum wage only if there is a risk of disrupting the local labour market, or to facilitate self-targeting.

The project must ensure that those who are unable to work, or unable to carry out hard labour, are not excluded. Some information must be gathered on how communities care for sick, heavily pregnant, elderly, disabled, and destitute members. Special consideration should be given to designing projects for socially and physically vulnerable groups (for example, elderly people, female-headed households, orphan-headed households); alternatively, other measures should be taken, such as providing vouchers or cash grants to meet the immediate food and non-food needs of the people who are unable to work. Some examples of this are shown in Box 9.

Community sensitisation and gender analysis should be carried out as part of the process of selecting the work activities. The gender analysis should include a consideration of control over household income, and the likelihood that women will be able to retain control over income earned, or influence the way in which additional household income is spent.

Household targeting is more appropriate than individual targeting in most cases, as this means that households with individuals who cannot work can also benefit. It is recommended that the project should employ one person per household. In communities where households are very large, or where a large amount of work needs to be done, or people have no other source of income, this limit may be raised to include more than one

---

Box 9: Care for vulnerable people who are unable to do heavy manual labour in CFW programmes

In **Somaliland**, those unable to work were provided with cash grants to enable them to meet their basic needs. Beneficiaries were selected through community-based targeting.

In **Kenya**, some communities allowed elderly beneficiaries to nominate younger relatives to do the work on their behalf. Destitute elderly women participated in a project that produced housing materials. They decided how much work they wanted to do, and when to do it. They were able to work at home and fit the labour around their other responsibilities. Elderly and frail people were able to participate, because the work was light and could be done from a seated position. Tasks such as project supervision, minding children, providing water, counting, and clerking were given to people who could not perform heavy labour. For women unable to work because they were caring for an elderly or sick relative, a collective system of care-giving in rotation enabled them to participate in the project.

In **Afghanistan**, CFW beneficiaries put aside a proportion of their wages to give to members of the community who were not able to work. Women unable to leave home for cultural or religious reasons participated in group textile production and kitchen gardening. Training was provided, and profits made from the textiles were ploughed back into the income-generating groups.

person per household. A system should always be agreed with the community to ensure that there is a fair distribution of work and consequently of income. If the work takes more than 15 days, consider rotating labourers after 15 days.

Setting pay rates

Labourers should earn at least enough money to meet their basic needs or basic daily living expenses. Ideally, project beneficiaries should be able to make enough money also to protect or recover their livelihoods. Where this is not possible (or necessary, if other forms of assistance are being provided), the project should state explicitly that the cash earned on the project contributes only to a proportion of immediate needs. Pay rates can be calculated in different ways:

- By comparing daily living expenses with current income sources.
- By comparing the current costs of basic food and non-food needs with the costs in 'normal' times.
- By comparing the costs of meeting basic food and non-food needs, and livelihood protection or recovery needs, with people's current income and assets.
- By using government information about the minimum or normal wage.
- By using local information about wage ranges for skilled and unskilled labour.

It is important to consult local government and other agencies about normal wage rates, and to persuade local NGOs, international NGOs, UN agencies, etc. who are working within the same geographical area to set common wage levels, in order to avoid disputes and conflict between communities. It is important to ensure that the 'normal' wage rates apply in an emergency context, when terms of trade and market prices may be inflated or deflated. Care should be taken not to set wages above normal labour rates if there is a risk that this will disrupt labour markets. Examples of pay rates for CFW are shown in Box 10.

Pay can be set per unit (cubic metres, number of acres, number of houses, etc.) of work completed and the number of people needed to do it; or it can be paid as a daily wage. Each community should agree on the hours that they wish to work per day, including starting times and timing of lunch breaks. In many cases, half a day's work is set to fit around a household's other essential activities. This ensures that women are not overloaded, that agricultural land and livestock are not neglected, and that people can still take advantage of opportunities for other, seasonal, employment.

Box 10: Examples of pay rates for CFW projects[1]

In **Kenya** in 2001/2, 70,000 beneficiaries received wages set according to activity. Rates of pay varied between 10,000 and 6500 Kenyan shillings[2] per kilometre of bush cleared. In Turkana, the project aimed to provide 10,000 Ksh/activity. In Wajir, units of work were determined for each project, and pay was determined according to how long this would take. For example, bush clearance was paid at 6500–8000 Kshs per kilometre, the preparation of *dufuls* (shelter material) at 600 Kshs per unit, and pan desilting at 200 per unit (2m x 2m x 0.5m).

In **Uganda** in 2001/2, the wage was set at 2500 Ugandan shillings[3] /day, which was a wage initially set by CARE International, taking into account the transport costs for beneficiaries and the number of work-days available. The average wage in the area for daily labour in normal times was 500–1000 shillings, but there was no report that the labour market had been distorted as a result of the higher pay.

In **Afghanistan** in 2003/4, wages reflected the market rate. In **Bangladesh** in 2001 and **Aceh, Indonesia** in 2005, rates of pay reflected the market rate for unskilled labour, and in **Matara, Sri Lanka**, CFW pay rates were specified by the government.

In **Ethiopia**, SC-UK has implemented a CFW programme over the past five years. It aims to meet basic food needs in bad years, and investment in livelihoods in better years. Payment rates followed guidelines set by the Ethiopian government of 5 Birr/person/day for 5 days/month. Every household member received assistance; so a household of five could earn 125 Birr[4]/month, for up to seven months. This doubled the annual income of target households. The rate of 5 Birr/person was based on the cost of 3 kg of cereal, which was previously earned in food-for-work programmes. When food prices increased to exceed 2 Birr/kg, local authorities had the option of switching from cash to food aid.

Pay by unit completed first needs a decision on how units of work should be measured. For example, when digging a water pan or dam, households may allocate a unit for clearing. A rate is then paid per unit cleared, based on an estimate of how long it will take. Road clearance might be paid per kilometre cleared. Materials collected might be paid for by volume (for example, bags of sand or numbers of sticks). The working units should reflect the amount of work that a team is able to undertake on a daily basis.

Removal of silt or earth can be measured in cubic metres or surface area: for example, *1.2 cubic metres per person per day*, or *one hectare per 20 working days*. Plantations of trees can be calculated in terms of preparations per day: for example, digging a hole, filling with it soil, planting a tree, filling in the hole, etc. Salt-pan reconstruction can be measured by using traditional steps: for example, construction of bunds, padding, trenching, sealing, filling, etc.

If it is difficult to divide a job into units, it will be necessary for beneficiaries and technicians to agree on the number of days that the job should take; payment should be made only for that number of days, regardless of whether the workers take longer. Tasks that cannot be divided up will need more teamwork and good supervision, to avoid disputes that may arise if certain beneficiaries perceive themselves to be working harder than others.

Management and staffing

Essential staff needed for CFW projects include field monitors, community supervisors, a logistician, and technical staff appropriate for the projects to be implemented. Some projects also need technical staff, including engineers, and food-security and livelihoods-support staff. It is important to budget for sufficient technical input.

The monitor is in charge of the payment register, which is separate from the attendance register. The number of monitors depends on how often they need to visit the communities, how much time needs to be spent in each one, and how mobile the monitors need to be. If CFW is implemented through local partners, then the number of agency-employed field monitors can be reduced, because the partner will bear the main responsibility for monitoring. In Bangladesh in 2004, one local partner employee supervised 200 beneficiaries, and the Oxfam GB field monitor needed to supervise only eight employees of the implementing agency on a project serving 1500 beneficiaries.

Each project should have its own community supervisor, elected by beneficiaries, to record attendance and work progress. A typical ratio is one community supervisor per 20 workers. In Haiti in 2004, on a canal-cleaning project in Cap-Haïtien, there were 10 workers per community supervisor. Community supervisors should be paid slightly more than the other beneficiaries, because they must attend work daily and are unable to share the workload with other members of their household. The community supervisor reports to the agency's field monitor or partner monitor, who is the key link with the programme manager.

A logistician will be needed, especially on construction projects. Timely delivery of basic tools and equipment is crucial. The logistician should be employed at the planning stage of the project, to allow for adequate preparation time. It is important for the logistician to be present at some of the project-identification or site-selection meetings. This will enable him or her to understand the type and quality of tools required for the project, thereby saving on the time needed for commodity requisition and delivery. The logistician's tasks include the following:

- Understand the requirements of projects proposed by the community.
- Investigate whether or not materials are available locally, to allow adequate time and avoid delays during implementation.
- Make fortnightly or monthly stocktaking visits, and store items near the project sites. Establish a sub-store; instigate and maintain an inventory system.
- Identify at the beginning of the programme the items that will be kept by the community members (hoes, shovels, baskets, etc.) when the programme is phased out, and items that will be returned to the agency. Regular inventory checks may be necessary to keep track of those items that will be returned to the agency.
- Ensure that procedures are in place to ensure a safe and secure working environment. For example, in Haiti supervisors received training from the Red Cross on the use of gloves and how to stem serious bleeding with first-aid pads. All injuries should be treated by community nurses or at the hospital, and all first-aid treatment should be paid for by the project.

In projects where there is a need for a skilled technician, contracts should be agreed for each job. Examples include spring protection and reservoir construction, where masons and other skilled workers are required. Where possible, local technical skills should be sought.

Transferring cash

The frequency of payments to beneficiaries depends on the objective of the programme. If programmes are intended to meet basic needs, payments will take place fairly frequently, in fairly small sums: for example, every third day in the early days of an acute emergency, with a gradual reduction to once every two weeks. At the beginning, it might be advisable to pay at shorter intervals until the community has gained confidence that the payments will take place. If the cash is intended to help people to recover their livelihood assets, payments should be large enough to do this, so they usually take place some time after the onset of the emergency: for example, after one month. Accessibility and logistics play a key role in the planning of payment days.

Planning the payments

It is important to maintain a register of the units of work completed by each beneficiary. The finance manager should check and approve the units and calculate the total amount of cash required for the project site. The programme accountant and programme manager double-check the sum of money allocated for each beneficiary and give their final approval before a cheque is prepared for submission to the bank. Other aspects of delivering cash and making payments are covered in Chapter 3.

Monitoring and evaluation

Monitoring should consider both process and impact. Baseline information is vital for meaningful monitoring and impact assessment. The general baseline information needed is indicated in Chapter 3. It is the same for cash grants and cash for work. Box 8 on page 56 gives additional economic information that needs to be collected to monitor the impact of a cash-for-work programme. Specific process indicators for monitoring the implementation of projects are listed below. In addition, certain impact indicators need to be monitored (and evaluated) specifically for CFW projects.

Process indicators

- Number of projects completed. Quality of the projects completed. Availability of technical expertise to supervise the projects.
- Were there enough people to do the work, and were they adequately skilled? If not, was relevant training provided?
- Was the working environment safe?
- How many labourers were employed?
- Were payments prompt, regular, and timely? Were the rates of pay appropriate?
- Number of labour days invested.
- Number of direct/indirect beneficiaries.
- Types of beneficiary (male/female, young/old, sedentary/mobile, possession of assets, type of livelihood).What measures were put in place to ensure equal participation by men and women?
- Has the project been able to promote equal and fair payment to all participants, regardless of sex and ethnic and social differences?

Impact indicators

- How has the project affected livelihood strategies? For example, labour migration, sale of harvest, borrowing money, investments in production, savings?

- Have beneficiaries been able to save some of their wages? Or repay debts?

- How did households manage the cash that they earned? Did it contribute to the food and income security of all household members?

- Are people economically active again, and utilising the assets provided by the CFW programme?

- What was the impact on the normal labour market and wage rates for casual labour?

- What was the impact on family relations, gender roles, etc.? Who controls the money? How did women cope with the workload?

- Were the projects themselves useful and relevant to the communities? What is the level of community ownership? How will the outputs be managed in the longer term?

- Alternative interventions: would beneficiaries have preferred cash grants, vouchers, food aid, food-for-work schemes, income-generating projects, etc.?

5 | How to implement voucher programmes

What are voucher interventions?

A voucher intervention aims to enable access to a specified range of commodities or services. Vouchers allow more choice than the direct distribution of certain commodities, but they can still be allocated to certain commodities.

There are two types of voucher. The first has a cash value, and it can be exchanged for a range of commodities up to that specified value. This type will be called a *cash voucher* in the remaining sections of this book. The second type is a voucher that can be exchanged for a fixed quantity of named commodities. This will be called a *commodity voucher*. Vouchers have been used to improve access to food, seeds, livestock, and other non-food items, and they have also been recommended to improve food utilisation: for example, to provide access to milling facilities in Darfur. Food vouchers are recommended only if the programme has specific nutritional objectives.

Vouchers are exchanged either with traders and retailers in shops, or with traders, middlemen, and local producers in local markets, distribution outlets, fairs, and other events organised for the purpose. Vouchers are appropriate in the following circumstances:

- There is a high risk associated with transporting and distributing cash.

- The market is weak and there is a risk of inflation and thus a need to encourage traders to move certain commodities into the affected area.

- The affected population identifies the need for specific commodities for which local supply is appropriate and available.

As explained in Chapter 2, the advantages of commodity vouchers are that they can protect beneficiaries from inflation, because people will always receive the same quantity of goods indicated on the voucher, regardless of the cost. When dealing with commodity vouchers, agreements with traders and shops need to be reached, in order to ensure consistent prices during the project period. In areas where inflation is rapidly affecting prices, agreements should be negotiated on a monthly or bi-monthly basis. Both types of voucher programme are relatively easy to monitor, because agency staff can check what the shop or trader has provided in exchange for vouchers, and what the beneficiaries have 'bought'. Voucher programmes are costly in terms of administration and financial overheads, because the vouchers have to be produced, coded, distributed, and tracked. It is important to allow time for these procedures when planning the project.

Sometimes relief agencies can provide specific commodities in exchange for vouchers. The ration card used in many food-distribution systems could in this sense be seen as a voucher for a fixed quantity of specific food items. This type of voucher system is not discussed in these guidelines, because it is essentially a form of direct distribution.

Food vouchers have sometimes been unpopular. Beneficiaries want to spend money according to their own priorities. If they have to spend it on food, they want to buy food according to cultural preferences and buy it where they choose, rather than in pre-assigned shops.

Box 11: Food vouchers and the dignity of beneficiaries

The use of food vouchers raises questions about the impact on recipients' dignity. Beneficiaries sometimes receive poor treatment in shops, and their choice is restricted. In Haiti, recipients of food vouchers reported that they would have preferred payments in cash, simply because they considered themselves capable of buying their own food: 'they preferred to buy a variety of food and to make their own financial decisions, and some of them felt humiliated receiving rice from the shops, like begging'.[1]

Similar conclusions are drawn in a report on the effects of a voucher scheme on asylum seekers in the UK,[2] which expresses reservations about the system because it negatively affected beneficiaries' dignity and, compared with cash grants, limited their choices and opportunities to buy the food and other essentials that they needed at prices that they could afford. In addition, having to buy food with vouchers was perceived by the asylum seekers to carry a social stigma.

Vouchers and fairs[3]

The fairs system, using cash vouchers, usually binds the relief aid to a specific sector (agriculture, livestock, etc.) but allows the beneficiaries to exercise their own preferences in terms of type, quality, and price of commodities. Vouchers may be exchanged at fairs that are purposely organised in areas affected by disaster. A fair is a space where traders can display their products, and buyers come to purchase what they need. The vouchers are the currency of the fair, but they have no value outside it. The fair is a place where producers and sellers can exchange information on the type, quality, and characteristics of the commodities on offer.

Fairs are usually organised when people are not easily able to obtain a specific commodity (seed, livestock, fishing tools, etc.), which is nevertheless available in sufficient quantities and quality within a reasonable distance of the affected area. Fairs have many advantages:

- The recipients can select from the commodities on display, and choose what best suits their needs.

- The system ensures a wide range of commodities.

- Fairs offer opportunities for social mobilisation and awareness-raising campaigns. In Zimbabwe, Oxfam GB facilitated HIV/AIDS education campaigns at fairs.

- The project is usually not responsible for managing the transport of the commodities (although in some cases, if travel costs are high, it may be necessary to subsidise the expenses of vulnerable households and producers).

- The project recipients are more active: they do not merely receive handouts, so they have a greater vested interest in the success of the scheme.

- Fairs give an opportunity to sell and buy products to the entire community. In Zimbabwe it was noted that after the 'official' seed fair the vendors stayed on and continued to sell seeds, thus increasing the availability of different varieties to the whole community, not solely to the project beneficiaries.

- They provide opportunities to exchange knowledge among buyers, producers, and traders.

- They mirror the 'normal' market trading system, ensuring a degree of dignity for beneficiaries while strengthening trading opportunities and links.

- Traders and local producers have access to cash, which boosts their businesses and their household economy.

Fairs can also play an important role in the management of local genetic resources. In voucher-fairs in Zimbabwe, farmers displayed seeds that many had forgotten, or that many had never known still existed. CRS conducted an evaluation of seed vouchers and fairs in Zimbabwe, Ethiopia, and Gambia, which confirmed the success of seed vouchers in providing seeds following a disaster. In all cases, enough seed was available to meet the needs of the affected population, despite initial assessments which sometimes concluded that there was not enough seed, as a result of production losses. There was a short-term positive impact on the area planted, but other benefits were also observed, in terms of re-building livelihood assets, strengthening local institutions and social relations, and reinforcing local seed systems.[4]

Box 12: Seed and livestock fairs in Haiti

In Haiti after the floods in 2004, Oxfam GB provided the most needy farmers with vouchers, which were used to purchase seed and livestock from a range of vendors (producers, middlemen, and traders) in fairs organised in disaster-affected communities. The fairs brought together farmers and merchants from neighbouring villages and gave local farmers the opportunity to choose from a range of varieties displayed. Voucher fairs permitted a quicker distribution of seed material, otherwise not possible through formal seed systems, and they offered an opportunity for farmers and traders to exchange information.

The varieties of seed and livestock exchanged were those most familiar to the local farmers. However, the fairs failed to display the expected wide variety of seed and livestock: dominant individuals put pressure on recipients to purchase from a limited number of traders, thus reducing recipients' free choice and creating some speculative price inflation.

Using local markets can provide an alternative to setting up fairs. Box 13 on page 72 illustrates Oxfam's experience in Niger.

Box 13: A market-based experiment in Niger

In 2005, Oxfam GB and its partner, AREN, implemented a voucher programme in Niger which linked food vouchers with weekly local markets. Beneficiaries receiving cash vouchers as grants or as payment for work activities were able to choose from among different food commodities displayed in local weekly markets. The project recipients could exchange vouchers though predetermined local traders, with whom Oxfam/AREN staff had worked to explain the system and the value of the vouchers. The vouchers mirrored the local currency (CFA), and recipients were permitted to receive small change in cash when appropriate. In contrast to the fairs system (discussed in the next section), this type of scheme makes use of the existing market and trading systems. The system enabled Oxfam to inform traders in advance every week about the demand, and thus the total value of the vouchers, and to subsidise traders to transport commodities in the most remote deserted areas. Agreements on prices were set with traders on a weekly basis. This market-based voucher system is not presented here in detail, because it is still being implemented at the time of writing.

Planning fair-and-voucher interventions

In addition to the elements common to other types of cash intervention, such as the forming of a relief committee and working closely with local authorities, there are specific tasks that need to be completed to organise a fair. Organising a fair can take as long as four weeks, especially when local partners are new to this kind of intervention. A reasonable number of beneficiaries per fair is somewhere between 400 and 600. More than 1000 would be difficult to manage.

In the Mapou project in Haiti in June 2004, the vouchers and fairs activities were implemented as a progressive process, initiated in a few communities and then replicated in others. The lessons learned from the first fair were documented, and appropriate measures were taken to improve the subsequent ones. The progressive process allowed more time to work with and build the capacity of the local relief committees.

Organising fair-and-voucher interventions involves the following steps.

- Identify sellers and suppliers.
- Mobilise the community, to help people to understand the system and to prevent problems arising during the fair.
- Disseminate information to buyers and sellers.
- Time the fair to coincide with local cropping calendars, in the case of agricultural inputs.

- Select beneficiaries according to the commodities specified on the vouchers.
- Select the location of the fair.
- Decide the date of the fair.
- Identify technical staff to check the quality of the commodities: for example, agricultural extension agents should check agricultural inputs.
- Set the value of the voucher.
- Produce the vouchers.
- Promote the fair.
- Identify opportunities to promote community awareness during the fairs: for example, the provision of information about HIV/AIDS, or gender issues.

The method for selecting the beneficiaries uses community-based targeting, as described in Chapter 3. When voucher-fairs are being planned, specific selection criteria may be set, depending on the objective of the intervention and therefore on the commodities to which the voucher provides access. Criteria for selecting beneficiaries might include (for agricultural inputs) access to land and labour availability; for artisan and trade goods, technical skills and trade activities; for orphan-headed households and households affected by HIV/AIDS, vouchers for the purchase of small ruminants (goats, sheep, etc.) might be appropriate.

In term of logistics and equipment, you will need some means of identifying team members (for example, T-shirts and caps with logos), sufficient and accurate scales, megaphones for making announcements, barriers for crowd control, and the necessary forms and documents. People to weigh goods and record transactions should be selected from the community or target groups. The venue should, if possible, be covered over, in case of rain or excessive heat.

The value of the vouchers depends on the objectives of the project, as well as the amounts and unit prices of the commodities that will be exchanged. Voucher-fair interventions usually aim to restore production and trade, or to re-establish productive assets. The value of the voucher depends on the level of production and assets that prevailed before the disaster, and the extent to which productive assets have been affected or lost. For example, the amount of seed needed should be the same as normal, determined by plot size, planting time, and product. Box 14 on page 74 illustrates how to calculate the value of a seed voucher.

Box 14: How to calculate the value of a seed voucher

Voucher value = price of the seed unit x amount of seed needed. The amount of seed can be calculated on the basis of the following information:

- The area of land that farmers can allocate for a specific crop; this includes land owned or rented by farmers, but also the human labour and animal draught-power available for land preparation and other cropping activities.

- The target yields for a certain crop, e.g. 750 kg of cereals for a household of five people, each consuming 150 kg per person per year.

- The seeding rate, which varies for different crops. Discussions with farmers will help to define the best seeding rate for each crop.

- The germination rates.

- The amount of seed that farmers are able to procure from their own production or other sources.

The increased volumes of supply and demand generated by the fairs can affect the initial prices of the commodities. In these cases, the project should ensure that the offer is able to satisfy the increased demand, in order to avoid price inflation and promote the presence of enough suppliers and producers to prevent monopolies operating. When the risks of inflation are high, the project should periodically negotiate agreements on prices with traders.

The location of the fair should be convenient for buyers and sellers, and at a reasonable walking distance from the recipients' households. Fairs may be organised in community areas, like schools or market spaces, but their boundaries should be well marked, easy to monitor, and large enough to accommodate buyers and sellers. The dates of the fairs should be agreed with communities and traders, to avoid clashes with other market days, religious festivities, community events, or activities.

Each voucher should include information about the beneficiary, and the place and date of the fair, to assist with monitoring and evaluation. They will be distributed on the day of the fair, and they should include a number of low-denomination vouchers, to make the exchange easier. To make the counting easier at the end of the day, each denomination should be identified by a different colour (perhaps the same colour as the currency notes used in the area, to reduce confusion). In Zimbabwe the vouchers were bound like a small chequebook, with each voucher dated and printed with a serial number. In Haiti, Oxfam distributed each set of vouchers in a transparent plastic bag, to make it easier to check their number and value.

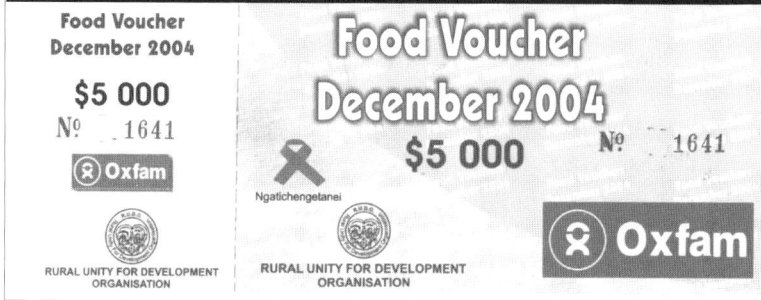

Figure 4: An example of a food voucher (Zimbabwe)

There are various ways to promote a fair: at key places such as markets and bus stations, using banners, leaflets, and megaphone messages; or via broadcasts from local radio stations. The information should include the date and place of the fair, and the types of product that will be exchanged. The sellers should be aware of the quality and variety of goods to be displayed at the fair.

It is important to ensure that information reaches local producers in the neighbouring villages, and that they have the means to transport their goods to the fair. In Haiti, sellers were mainly middlemen rather than local producers, a fact that limited the range of the goods displayed and the exchange of information among sellers and buyers; also it had the effect of slightly inflating prices. Spreading information to local producers and traders, informing them about the number and total value of the vouchers that will be exchanged during the fair, will encourage their participation.

It is important to consider the costs to the beneficiaries of carrying the product back to the community: consider offering subsidies or assistance with transport if distances or terrain pose problems.

On the day of the fair

On the day of the fair, the following procedure should be followed.

- Sellers should arrive early in the morning at a specified time.
- At the entrance to the fair, registration teams should weigh the commodities and record the amounts and types of items for sale. A specific time should be allocated for this stage (two hours, for example), and more than one registration team (at least three teams of two persons) should work at the same time, to speed up the process.
- At least two representatives of the relief committee should systematically inspect the quality of the commodities. In some cases they should have technical expertise: for example, agricultural extension agents should check agricultural inputs.

- When the registration and inspection of commodities is completed, distribute the vouchers to the beneficiaries (who should arrive at a specified time).
- Invite the beneficiaries to tour the fair, without any exchange taking place.
- Announce the prices and remind the beneficiaries and traders about the process to be followed.
- Then the fair itself can start.
- When the exchanges are finalised, record the number and value of vouchers per seller, for reference when payments are made later.

These phases should be kept separate from each other, in order to avoid confusion and to monitor the process of the fair efficiently.

To ensure the visibility of the team and the relief committee, it is a good idea to identify those involved in specific tasks such as registering the sellers, distributing the vouchers, and checking the quality of goods, by giving them colourful printed T-shirts and hats/caps to wear.

Allow enough time to remind everybody about the process and purpose of the fair. This will reduce the risk of abuse or cheating. It is important to monitor prices, to detect any attempt by sellers to fix prices or speculate on them. The higher the number of sellers participating in the fair, and the wider the range of goods on offer, the lower the risk of price inflation. Consider imposing a limit on the volume of goods that each trader can display and sell during a fair, to prevent powerful traders operating a monopoly in the market.

Monitors should be present to assist buyers and sellers in the use of the vouchers. They should explain that the vouchers can be used only in that specific fair; but that beneficiaries are permitted to use their own money to increase the purchasing power of the vouchers at the fair.

At the end of the fair, sellers should redeem the vouchers for money, paid out by the relief committee under supervision by agency staff. Payment may be made either in cash or by cheque, depending on the context. To enable payment to be made on the day of the fair, the event must end at least two hours before sunset. Otherwise it is advisable to arrange for sellers to be paid on another day.

The fair is concluded by recording the amount of goods that have not been sold, in order to evaluate the types and quantities that have been exchanged and preferred by the beneficiaries. The amount sold could exceed the value of the vouchers exchanged, because buyers may have used their own money to purchase additional goods.

Monitoring and evaluating fairs

Chapter 3 describes the process of monitoring and evaluating the distribution of cash grants. Below we consider aspects that are specific to voucher and fair interventions. The registration process in itself will provide information about the numbers and types of participants, including sex, age, and origin of traders, producers, and recipients. The registration will also provide information on the volume of commodities exchanged.

At the end of the fair, questionnaires should be used with a sample of buyers and sellers, to assess their perceptions of the following factors.

- What went well and what went badly?
- Was the targeting process transparent and fair?
- Were any vouchers sold to others?
- Were the commodities of good quality?
- Were people able to purchase what they were looking for?
- Were the prices higher or lower than at the local market?
- Were people able to exchange information about the commodities on offer?
- Were the vouchers used correctly?
- Were the date and time appropriate? Was sufficient advance notice given?
- Would the beneficiaries have preferred another type of intervention, such as cash grants or direct inputs?
- Will they be able to use all of the products that they obtained at the fair, or did they sell or barter some of them to meet other needs?
- How far did buyers and sellers have to walk to reach the fair?
- Was the payment to sellers made in good time?
- Was the timing of the fair appropriate in terms of seasonal activities, such as sowing and planting?
- Was the value of the voucher adequate in relation to the prices of commodities?
- Did any accident occur? If so, what impact did it have on the fair?

Management and staffing for fairs

Oxfam's experience of organising fairs in Zimbabwe and Haiti suggests that between 10 and 12 persons should be sufficient to manage the process. The team should include monitors from the agency and local implementing partners. To strengthen community empowerment, the major part of the monitoring team should comprise representatives of the relief/fair committee and staff from the local ministries.

The team and the committee should closely monitor the fair itself, making sure in particular that nobody is exploited by sellers who might take advantage of beneficiaries who are not familiar with the voucher system. In paying vendors, Oxfam's Zimbabwe programme used staff from the finance department in the head office, who travelled out from Harare a few days after the fair. This ensured that the accounts and payments were rigorously monitored.

Vouchers and shops

In addition to fairs, a voucher intervention can also be implemented through local shops. The shop system may utilise either cash vouchers or commodities vouchers. The beneficiaries come to the shop and collect goods upon presentation of their vouchers. The items may be collected at any time within a specific period. It is helpful to set specific days for participants to collect items, in order to facilitate book-keeping and monitoring processes. The shops may sell food and/or a number of other items. The vouchers may allow beneficiaries to decide what to buy from a range of specific goods, or they may be tied to specific commodities.

The following practical steps refer mainly to Oxfam GB experience in an urban context in Haiti. In Cap-Haïtien, Oxfam set up a voucher system through community shops, to manage payments in cash and in kind for work on a canal-cleaning project. In an urban context characterised by high levels of insecurity, 2000 beneficiaries were paid with vouchers in 50 community shops every fortnight (1000 at a time, on Fridays). In the shops they received part of their remuneration in cash and part in kind (rice).

The main reasons for adopting a vouchers and shops system are as follows:

- To enable local shops to be involved in the life of community.
- To provide a cash boost to small shops and the local economy, rather than going through a bank.
- To reduce the risk of non-payment, since the shops are based in the community.
- To limit risks to security, since each shop pays a maximum of 20 beneficiaries each week and so manages only a small amount of cash.
- To minimise the necessary logistical support.

In urban and insecure contexts, a voucher programme has distinct administrative and security-related advantages over commodity distributions, for the following reasons.

- Only a few bank cheques are drawn each week.

- Direct management is much easier: in previous commodity distributions, trucks have been looted en route to distribution points, and there has been insecurity at the distribution points.

- A voucher scheme is a self-controlling system. In Haiti it was easily understood by participants, who often resolved problems (such as a beneficiary trying to exchange vouchers at two shops in the same week) without Oxfam's intervention.

- Once set up, the management system is easy to use and very effective.

- Agency staff do not handle cash.

Planning a shop-and-voucher intervention

In broad terms, the stages in the process of implementing a shop-and-voucher system are as follows:

- selecting local shops

- preparing vouchers and ID documents

- distributing ID cards and weekly vouchers to beneficiaries

- distributing weekly forms to selected shops.

Selecting local shops requires careful identification of the merchants, based on assessments of their respectability, literacy and numeracy, reliability, and length of time established in the community. It is important that vouchers' recipients live close to the shops, in order to limit any possible rivalry between neighbourhoods and to ensure that recipients will regularly exchange the vouchers, at prescribed times. Selected shops need to be well known and accepted by the community.

The ID cards and weekly vouchers should include the following information: *name, address, name of shop, address of shop, the amount of each commodity to be received, date, name of team leader, name of supervisor.* Each of these should be coded, to assist with control of the system. Controlling and monitoring the system is explained below under 'Monitoring and evaluation'.

The community shops should receive a weekly table that includes the following information: *name and address of shop, list of beneficiaries for that week, quantity of each commodity due to each beneficiary, date, name of supervisor* (in Haiti, one supervisor was allocated to each shop). Each category of information should be coded.

DFID Department for International Development

Oxfam

IDANTIFIKASYON (Identification)
N⁰

Siyati (Family name) :   **Non** (first name):

Sèks (sex): **Kategori** (category):   **Kòd Ekip** (team code):

Kat sa a ap bout jouk 30 Sektanm 2004 – Valid up to 30 September 2004

The community shops also receive a bank cheque, amounting to the total value of the commodities to be distributed, together with an additional amount corresponding to their remuneration on a weekly basis. The shops' compensation should be set very carefully; when applicable, it should not be higher than the local bank's commission fees. Excessively large rewards to shops may cause suspicion among beneficiaries, who may accuse the merchants of exploiting the system.

Management and staffing of a shop-and-voucher system

The management system necessary to run a shop-and-voucher scheme should not be underestimated. Close monitoring is required on payment day, to ensure that the quantity and quality of the commodities are appropriate and that the vouchers system is not abused. In Haiti, two Oxfam Project Officers were able to monitor 20 shops through 20 supervisors for a total of 2000 beneficiaries. Management also requires a payroll-system manager with high computer literacy and close attention to detail; this person's task is to swiftly enter data about daily work attendance (in the case of CFW) in order to meet the payment deadlines, prepare payment vouchers, and reconcile actual voucher-based payments with advances given to the shops.

Monitoring and evaluation of a shop-and-voucher system

Several aspects of a shop-and-voucher system need to be monitored. They are very similar to the monitoring of a commodity distribution. They include monitoring the distribution process itself, which consists of on-site monitoring while the distribution is carried out and checking the accuracy of the distribution through the control system, as explained in Figure 6; and monitoring the use and impact of commodities received at the household level.

Figure 6: Example of a shops control system

STEP 1

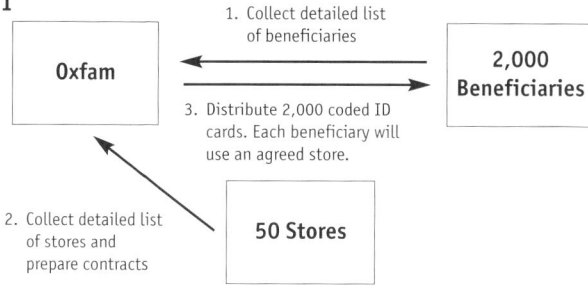

| Oxfam | 1. Collect detailed list of beneficiaries | 2,000 Beneficiaries |

3. Distribute 2,000 coded ID cards. Each beneficiary will use an agreed store.

2. Collect detailed list of stores and prepare contracts

50 Stores

STEP 2

Oxfam

1. Distribute food vouchers and cash vouchers

2,000 Beneficiaries

2. A cheque is given to stores for exact number of cash vouchers and food vouchers (this will include a handling fee amounting to 25%)

50 Stores

3. The stores purchase the required quantities of rice to distribute on basis of projected number of food vouchers. These will also prepare the cash

STEP 3

Oxfam

1. Beneficiaries bring food vouchers and cash vouchers to stores

2,000 Beneficiaries

50 Stores

2. Rice and cash distributed to 2,000 beneficiaries

STEP 4

Oxfam

1. Food vouchers and cash vouchers collected by Oxfam

2,000 Beneficiaries

50 Stores

The control system works as follows. The beneficiaries hand over their vouchers to the shops, in exchange for commodities. The shop owner checks the information on the vouchers, ensuring that it matches the details on the ID card and the weekly table. Then the owner distributes the items. Agency project supervisors then collect all the vouchers from the shops. These vouchers are entered on a database, using the relevant codes to speed up the process. The database is then used to reconcile the bank cheques given to the community shops and the vouchers that they collected from the beneficiaries. Shop owners must understand that a single instance of misappropriation of funds would result in permanent exclusion from the project.

Household monitoring, or end-use monitoring, involves selecting a random number of beneficiary households for questioning about their involvement in project activities, their use of vouchers, the amount received, the security of the system, their degree of satisfaction with the system, and the impact of the scheme on their household income and expenses. The following set of conditions describes an ideal shop-and-voucher programme:

- The shopkeepers supplied the agreed products, of an appropriate quality and quantity, to the beneficiaries without any dishonest practice.

- The process was transparent, and the community was able to monitor it, because everyone had been informed about the value of vouchers, and the products against which they were redeemable.

- The voucher was appropriate, and the community was not forced to sell a high proportion of the goods to meet other needs.

- The beneficiaries had been effectively targeted and were the most vulnerable members of the community.

- There were no risks to the safety of the beneficiaries when they exchanged and handled the vouchers/commodities.

- The beneficiaries were satisfied with the cash/commodities that they received, and they were not mistreated by the shopkeepers.

The monitoring and evaluation should also include a question about the impact of the voucher system on the local economy at the level of shops and households; these concepts are discussed in Chapter 3.

Conclusion

This book has argued that in many emergencies, the appropriate response to meet people's food and non-food needs is to give them the cash to buy these items themselves. This is a faster and more appropriate way of meeting people's needs. It also maintains the dignity of the disaster-affected population and empowers communities to prioritise their needs. Furthermore, cash interventions can boost and revitalise the local economy.

The book shows how to implement cash programmes, based on more than five years of Oxfam experience. Yet, despite the fact that this type of intervention is not new, cash-transfer programmes as an emergency response are still relatively rare. So what is stopping donors from funding cash-based programmes, and humanitarian agencies from implementing them, more often? The main deterrents seem to be fears that the cash will be misused or diverted, or pose a threat to security.

The challenge for humanitarian agencies in the coming years is to overcome these fears, and to implement cash programmes where they are judged to be the most appropriate response. We need to find a way of trying out new, more, and bigger cash programmes. This will require a change in donor policies, but also a fundamental change in the mind-set of aid workers, who are used to identifying needs and providing commodities and thus to maintaining control over the provision of assistance. One consequence of cash programming is that agencies must hand over control of many aspects of emergency programming to the affected communities themselves.

The unprecedented public response to appeals for money to help survivors of the Indian Ocean tsunami has allowed us to do this. It has already led to the implementation of many different types of cash programme by Oxfam and others. We need to carefully record and monitor what we do over the coming years: to improve our programming by finding out what works and what doesn't. As we gather evidence of the success of cash programmes, and examples of how to implement them, agencies will become more confident to try them. Success stories will also give powerful justification for changes in the policies of donors.

Cash transfers are appropriate to meet emergency needs and to rehabilitate livelihoods, but for many population groups they are unlikely to be sufficient on their own, and unlikely to lead to long-term sustainable livelihoods. Many people suffer prolonged or repeated emergencies. In theory, emergency cash transfers should continue until people are able to meet their minimum requirements through their own means, or national relief or social protection systems can take over responsibility. For most poor and emergency-affected populations, such aims will seem far from their day-to-day reality. Recent renewed interest on the part of donors in social protection schemes, however, makes the establishment of long-term financial support for the poorest a real possibility. The possibility of longer-term social protection programmes and the potential to link them with emergency cash programmes also creates new opportunities to link emergency and development programmes.

In the current context, where we have both the commitment and the means to find new and better ways of meeting the needs of the poorest and the most vulnerable, we should take every available opportunity to try out innovative programmes. We have a crucial opportunity to change the way in which humanitarian assistance is provided – and we need to take it.

Appendices

Appendix 1 | Logical Framework (Oxfam response to Mapou floods, Haiti, 2004)

LOGICAL FRAMEWORK

	Intervention Logic	Objectively verifiable indicators of achievement	Sources and means of verification	Assumptions and risks
Overall objectives	To save lives and protect livelihoods of flood-affected communities of Mapou, in south-east Haiti			
Project purpose	To improve food security and to ensure economic incomes of flood-affected communities in Mapou, in particular women and vulnerable households, by re-establishing productive assets and access to market, through cash and vouchers transfer mechanisms.	Vulnerable households have enough income and produce to meet their food needs	Food-security assessment baselines Feedback from community-based committees	No adverse climatic factors affecting the communities during the project period
		Levels of malnutrition are controlled	Secondary data, other agencies reports	Cooperation among agencies, local government, and partners
		Reduction of distress-coping strategies among the affected population	Programme activities and monitoring reports	National political and economic environment remains stable
Expected results	1. Vulnerable farming households restore their agricultural and livestock assets and their production capacity through the provision of seed and animal vouchers and the organisation of local fairs.	Seeds and livestock are obtained through existing in-country supply systems and satisfy farmers' preferences.	Other agencies' reports (especially FAO)	No adverse climatic factors constrain agricultural and livestock production
		Timeliness: seed fairs take place within seeding season (August)	Field observations	Security for cash management
		300 farmers purchase seeds at the fairs	Fairs monitoring	Access to the population affected by floods
		Number of vendors (farmers, small traders) selling seeds in the fairs	Crop-cycle monitoring	Funding is available to allow timely intervention
		Number and type of crop varieties displayed and exchanged during the seed fairs	Post-distribution interviews with beneficiaries	
		90% of seeds bought at the fairs are sown by farmers	Feedback from community-based committees	
		300 affected farmers have restored at least 50% of their normal crop production	Project reports and monitoring	
		300 farmers purchase livestock during fairs; at least 75 are women		No outbreaks of epidemic disease among local livestock

				head of family
				Stability of local and national livestock market prices
		Number and species of livestock displayed and exchanged at the fairs		
		Livestock sold are healthy and satisfy farmers' preferences		
		Number of fairs organised and number of vendors selling livestock		
		Distance that farmers have to walk to reach the fairs		
		300 households have restored their livestock assets		
Expected results				
	2. Women, particularly heads of families, restore their business and petty trades activities through the provision of income-generation packages.	175 women are able to restart their business activities	Focus-group discussions with women; household interviews	Access to market
		175 productive packages distributed	Post-distribution monitoring	Market supply chain is functioning
		Number and type of business activities supported	Project activities report and monitoring	Local partners are available
		Volume of business that beneficiaries are able to manage		
	3. Economic income assured for 200 affected households through their participation in cash for work schemes.	Percentage of the affected population is involved in CFW activities	Cash-distribution report	Markets are functional and food is available
		200 men and women involved in CFW activities receive a remuneration adequate to their needs	CFW group leaders' reports	No price inflation or currency devaluation
		200 CFW beneficiaries receive timely and regular remuneration on a weekly basis	Project reports and monitoring	Security is assured in the work places
				Cash can be safely managed
	4. Wider communities benefit from the	CFW projects are selected by the communities	Feedback from community-based committees	No climatic change will delay community work activities

continued ...

public work schemes, which could include construction of reservoirs, rehabilitation of access roads, and other small-scale projects selected by the same communities.

CFW projects are feasible and environment-friendly	The CFW community projects are completed by the end of the project	Project works monitoring

Means:	**Costs:**
Personnel	
Travel and subsistence	
Office and running costs	
Transport	
Micro-enterprise productive packages	
Seeds and livestock vouchers	
Cash-for-work payments	
Cash-for-work tools	
Monitoring	

Activities

1. Recruit staff
2. Collect baseline information
3. Establish community-based committee (including men and women) and revise selection criteria
4. Inform and sensitise local seed and livestock sellers about fairs
5. Organise seed and livestock fairs
6. Monitor seed quality and crop performance
6. Set up business plans with women
7. Distribute productive packages to women
8. Community-based identification of the CFW projects
9. Technical design of the CFW project
10. Community-based targeting of CFW participants
11. Implementation of CFW projects
12. Establishment of alliances with local groups and organizations
13. Monitoring

Appendix 2 | An attendance sheet for a cash-for-work programme

UNSKILLED WORKERS

Type of Cash-for-Work Activity: _____

Name of Field Supervisor: _____

Location: _____

#	Name, Surname of CfW beneficiary	Worker details				Dates of CFW activities					
		Age	Sex	ID no. (if available)	Address	14/02/05		Total		Total	Total
1	Example										
2											
3											
4											
5											
6											
7											
8											
9											
10											
	Total										

Appendix 3 | A payment requisition form for a cash-for-work programme

Oxfam

Requisition for CFW Payment
(one per week per project)

Date: ..../......./2005

Week from .... To ....

Type of CFW activity: ....

#	Name, Surname of CFW beneficiary	no. working days	Daily rate	Total	Breakdown of Total in available banknotes (for finance staff only)					
					1000	500	100	50	20	10
1	Example	3	300	900	0	1	4	0	0	0
2										
3										
4										
5										
6										
7										
8										
9										
10										
	TOTAL			0	0	1	4	0	0	0

Requester:
(name and signature)

Received by (OXFAM assistant):
(name and signature)

Budget holder approval:
(name and signature)

Appendix 4 | A daily payment sheet for a cash-for-work programme

Cash For Work Payment Sheet
(1 per week per project)

Oxfam

Date: ....../....../2005   Type of CFW activity: ....

Payment for week from..............to...........

#	Name, Surname of CfW beneficiary	# of work days	Daily rate	Total	Received (beneficiary signature)
1					
2					
3					
4					
5					
6					
7					
8					
9					
10					

Paid by (Oxfam LLH assistant):
(name and signature)

Witnessed by (Supervisor)
(name and signature)

summary statement of float (to be filled by Oxfam LLH assistant)

amount taken:
total paid:
refund:

date:
(name and signature)

FA:
(name and signature)

Appendix 5 | The market-supply chain – example from Haiti

In Haiti, the market analysis tool was used to assess the impact of the floods on the market as a whole, and to determine possible cash-transfer interventions to re-establish the initial market-supply chain. The analysis began by assessing the value chain before and after the floods. The first step was to identify the actors who were trading key foods and non-food items considered essential for survival and for livelihoods. The key actors were then interviewed; they included local consumers, the women who act as transporters (Madame Saras) between villagers and middlemen, and retailers. The results were combined with information gathered from farmers' organisations and local community-based organisations.

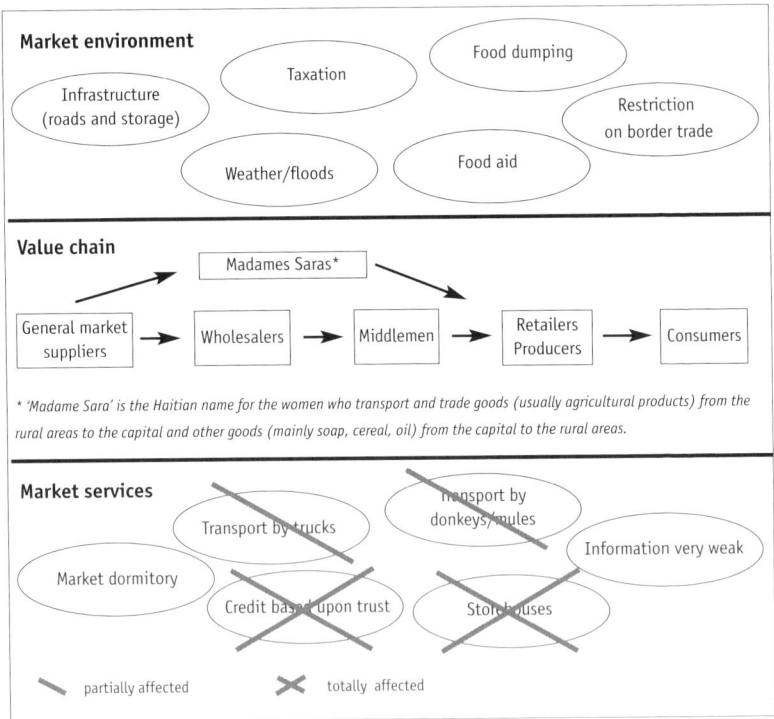

Market environment
- Infrastructure (roads and storage)
- Taxation
- Food dumping
- Weather/floods
- Food aid
- Restriction on border trade

Value chain

General market suppliers → Wholesalers → Middlemen → Retailers Producers → Consumers

Madames Saras*

* 'Madame Sara' is the Haitian name for the women who transport and trade goods (usually agricultural products) from the rural areas to the capital and other goods (mainly soap, cereal, oil) from the capital to the rural areas.

Market services
- Market dormitory
- Transport by trucks
- Credit based upon trust
- Transport by donkeys/mules
- Storehouses
- Information very weak

partially affected   totally affected

In the flood-affected areas of Haiti, the main staple food and other primary commodities were supplied by a few major local wholesalers, who purchased goods directly from Port au Prince, getting zero-interest loans from general market suppliers on the basis of acquaintance and trust. The wholesalers supplied goods (rice, sugar, flour, oil, beans, cement, etc.) to middlemen, who usually had limited transport facilities, such as donkeys and mules. The middlemen sold commodities in small amounts to numerous retailers, who took the goods on a daily credit basis and sold them in the more marginal areas. Alternatively, Madame Saras would buy direct from general market suppliers in the capital city and supply the retailers. In some cases, producers sold their commodity directly in local markets.

As a result of the floods, wholesalers lost their transport and storage facilities, because trucks were damaged and storehouses were destroyed. They had been left with debts to pay, and it was impossible for them at that point to obtain further credit. Middlemen and retailers, including Madame Saras, had been affected both in terms of transport (through loss of their pack animals) and in the loss of the stocks that they kept for sale. The consumers lost both assets and income-earning opportunities, and their purchasing power was therefore much reduced. The general market suppliers were not affected.

Oxfam's response to the floods focused on restoring the far end of the supply chain, targeting the most vulnerable groups: the poorest consumers (including wage labourers), producers, middlemen, and retailers. Oxfam also re-established some of the market services disrupted during the flooding, such as transport, access to credit, and market information. Cash interventions were considered appropriate, because those most affected had lost assets and income, and – with assistance for Madame Saras – local markets could be supplied with the necessary goods. Assistance to wholesalers was not considered necessary, because these were among the wealthiest in the community. Hence the cash interventions were targeted at the following:

- Consumers: through employment activities (CFW) that increased the purchasing power of 500 vulnerable households.

- Retailers and Madame Saras: in particular women selling commodities in the local markets, by giving them vouchers to purchase the basic commodities to re-start their petty trade activities.

- Middlemen and producers: through organisation of vouchers and fairs, where producers and middlemen had the opportunity to display and exchange seed and livestock, and producers had the opportunity to buy them.

- Transport: by rehabilitating roads that connected affected communities with local markets, as part of the CFW activities.

- Market information: fairs and vouchers promoted the exchange of information among producers and traders about prices, types, and features of the commodities available.

Appendix 6 | An information sheet for a cash-for-work programme (Kenya, 2001)

Oxfam GB, Turkana - Recovery Programme, November 2001

The Oxfam Recovery Programme in Turkana intends to introduce a programme of cash-for-work (CFW) projects in Lokitaung sub-district. The programme has three aims:

a. To provide support to drought-affected communities through a direct cash transfer.

b. To select appropriate and technically viable work projects that will benefit the wider community.

c. To contribute to Oxfam's understanding of alternatives to free food distribution and Oxfam's understanding of the impact of CFW.

Why cash for work?

Food relief may save lives, but it only solves the immediate problem. If pastoralists are to recover from drought, further resources are required which will enable them to rebuild their livelihoods, pay for essential basic services, and thus recover rapidly from the effects of the drought.

Advantages of cash for work

- It provides households with a degree of choice with regard to their own spending priorities.

- Cost-effective in comparison with alternatives (restocking, seed distribution, etc.).

- Low distribution costs.

- Beneficiaries receive greater proportion of donated money.

- Boosts the local economy.

- The CFW projects themselves provide social benefits to the community as a whole.

- Can improve the status of women and marginalised groups.

- Self-targeting, because wages will be at an unattractive minimum.

- Reduces risk of corruption (money is earned and hence is more valuable to beneficiaries).

Disadvantages of cash for work

- Work is often inappropriate for the most vulnerable (sick, old, children).

- The workload of women is increased.

- Women may not retain control of income.

- Cash may be abused, e.g. by the purchase of tobacco and alcohol.

- Only viable in cash economies.

- Higher security risk.

- Inflationary pressure.

- May affect community participation in future community projects.

- Potential for corruption.

Measures to reduce disadvantages

- Community sensitisation and training in effective methods of managing cash.
- Clear criteria for selection of beneficiaries.
- Ask communities for solutions to assisting most vulnerable.
- Extra security measures and close monitoring of payments.
- If cash is not commonly used, give out goods for barter.

Projects

- Communities have been asked to suggest projects/activities that they believe will improve their lives.
- Other stakeholders (government and NGOs) are being consulted as part of the selection process.
- Oxfam staff and the relevant government departments will provide technical support.
- Proposed projects include de-silting or digging pans, building reservoirs, shallow wells, and troughs, and clearing roads. If communities have other ideas, they will be considered.
- Lasting impact and sustainability will be key criteria in the selection of projects. Communities must see the potential benefits of the project and therefore have an interest in making the project succeed.

Beneficiaries

- It will be the responsibility of the Relief Committee and the community to select beneficiaries.
- Selection will be based on criteria outlined by Oxfam but clearly defined by the community through discussions of what constitutes vulnerability in their community.
- Beneficiaries must be the poorer members of the community: those unlikely to recover from the effects of drought without assistance (i.e. households with no animals).
- Women will be targeted in order to improve the nutritional status of families.
- Possible solutions to targeting the most vulnerable (sick, old, and disabled people) will be discussed with the whole community; for example, payment per task, for increased flexibility; and selecting whole households as beneficiaries, to enable different members of the family to contribute.
- Screening of beneficiaries will be done with the involvement of the beneficiary community.

Measuring impact

- Recipients' use of cash, compared with use of income over past two years (data supplied by food-security assessments).
- Market prices will be monitored, and traders will be interviewed.
- The extent to which the most vulnerable members of the community have been able to participate in CFW; to be assessed though discussions with the community.
- Movement of people into CFW areas will be monitored.
- Number of CFW community projects started and successfully completed and maintained.

Appendix 7 | Community-based targeting[1]

The following steps are normally followed in community-based targeting:

1. Implementing agency holds public meetings with local authorities and community members to explain the nature of the programme. If only a proportion of the population is to be targeted for assistance, this is also explained.

2. The community elects a relief committee (RC) at a public meeting. The aim should be to have a broadly representative committee, including adequate representation of women, and the ethnic, social, economic and political groups within the community.

3. The implementing agency and RC discuss the criteria that should be used for beneficiary selection. These criteria may then be discussed in a public meeting.

4. The RC registers the beneficiaries for the programme. The list of beneficiaries may be read out at a public meeting, so that everyone in the community has a voice in the process of ensuring that only the poorest or worst-affected, and all those meeting the selection criteria, are included in the programme.

5. Distribution, or payment, is carried out by the RC, together with a monitor from the implementing agency.

6. The RC receives feedback on the targeting and distribution method from the community, and informs the agency monitor. The monitor arranges for the programme to be adapted, if necessary, to be more effective. The community can also give direct feedback to the monitor on any issues regarding the relief committee. New members can be elected if some are not found to be fulfilling their responsibilities.

7. Post-distribution monitoring is carried out by the implementing agency, with input from the RC.

Community based targeting has been found to be most effective in the following circumstances:

*   Conditions are relatively stable, with no acute conflict.

*   There are identifiable differences within communities, for example large wealth differentials between those targeted and not targeted.

*   A fairly large proportion of the population is targeted for assistance.

*   The community co-operates with the targeting process.

*   Community representatives are accountable, so that distributions do not risk large-scale diversion.

*   The supply of resources available broadly matches the immediate needs of the population.

In the absence of these conditions, the targeting system is likely to become very expensive, due to increased staff requirements, or will have a wide margin of error.

Appendix 8 | A monitoring form for a seed fair (Zimbabwe)

Key indicators	Sources of information	Purpose/objectives of gathering this specific information
Number of targeted people per ward vs. total population Category of beneficiaries (orphans, widows, people living with HIV/AIDS, etc.) Dates of distribution % of beneficiaries involved in previous distributions	Village Relief Committee (VRC), fairs, counsellors, communities	To maintain records of the scale and scope of coverage, for quality reporting, accountability, impact assessment, and future planning
Type/amount of seeds sold Average prices/kg/seed Type/number of vouchers Amount spent per ward	Fairs/traders	To check if implementation accords with the plan, and if the recommended minimum packages of seeds have been received by the beneficiaries. To determine the reasons if this was not achieved
Sources of the seeds (local or external?) Number of traders Estimated amount of seeds in the market	Fairs, traders	To understand the extent to which the local market was capable of meeting the need for seeds, and to determine what proportion of the fund for the seeds stayed in the local economy
Mechanisms by which the prices were agreed	Fairs, traders	To check if prices at the fairs reflect local prices. To determine impact of the fairs on the local market
Quality control	Ministry of Agriculture	To determine whether quality control has been done by the responsible agencies
Roles of various bodies (e.g. government, Oxfam GB; including mainstreaming of HIV/AIDS and gender)	District Dev. Fund, local councillors, communities	To determine the level of co-ordination To identify challenges and future action points and opportunities
Method of targeting Criteria used for selection Effectiveness of communication	VRC, communities	To document the targeting methods practised, lessons learned, and challenges faced
Gender composition Effectiveness and transparency of VRC VRC's length of service	VRC, beneficiaries, local leaders, councillors	To determine if VRC adequately represents the community, to assess the effectiveness of the VRC in co-ordination, and to identify capacity-support requirement
Typical comments/views on targeting (fair, not fair?) and the level of attendance (large, medium, small)	Women- and orphan-headed households; those missed by the targeting; people living with AIDS	To get feedback from communities on the effectiveness of the targeting in reaching the intended beneficiaries
Distance travelled to the fairs by many of the households	VRC, communities	To check if beneficiaries travel long distances. To take corrective actions for subsequent activities
Proportion of seeds resold in the market	Market/key informants	To determine the likely effectiveness of the distribution
Other lessons, including feedback from communities	All sources	Get feedback on any key issues, challenges, opportunities

Notes

Introduction

1 'Minimum food and non-food needs' are defined in the Sphere Project's Minimum Standards in Disaster Response. Depending on circumstances, 'non-food needs' might include hygiene items, clothing, kitchen utensils, shelter materials, or health services. The term also includes livelihood assets such as seeds, tools, livestock, fishing equipment, and transport.

2 The Sphere Project: *Humanitarian Charter and Minimum Standards in Disaster Response* (Oxford: Oxfam Publishing, 2004).

Chapter 1

1 Sen (1981)

2 Sperling and Longley 2002

3 Khogali 2001; Bonal 2003; Frize 2002; Jones 2004; Creti 2005; Khogali and Takhar 2001.

4 Khogali and Takhar 2001; Jones 2002; Creti 2005.

5 Lothike 2005.

6 Brocklebank 2005.

7 Frize 2002.

8 Adams and Kebete 2005.

9 Buchanan-Smith and Borton 1999.

10 Adams and Kebete 2005.

11 Levine and Chastre, 2004, July.

12 Acacia Consultants 2003; Brocklebank, 2005.

13 Buchanan-Smith and Barton 1999; Khogali 2001; Khogali and Takhar 2001; Frize 2002; Jones 2004; and Creti 2005.

Chapter 2

1 Examples of shocks: a flood, earthquake, drought, conflict, and economic collapse.

Chapter 3

1 Ali *et al.* 2005.

2 *Ibid.*

3 £1 = between 175 and 200 Sri Lanka Rupees (2004/5).

4 GTZ, Lusaka 2005.

5 SDC and UNHCR 2002.

6 Adams 2005.

7 Harvey 2005.

8 Horn Relief, Novib, and Oxfam GB 2005.

9 Hofmann 2005.

10 Jones 2004.

Chapter 4

1 Sources: Frize 2002; Khogali and Takhar 2001; Jones 2004; Adams and Kebete 2005.

2 £1 = ~ 114 Kshs (2000/01).

3 £1 = ~ 2,400–2,500 Uganda Shillings (2000/2001).

4 £1 = 12–16 Ethiopia Birrs (2000–2005).

Chapter 5

1 Hadcroft 2004.

2 Oxfam GB, TGWU, and Refugee Council 2000.

3 Information for this section is based on Oxfam experiences in Haiti and Zimbabwe; guidelines produced by ICRISAT (International Crop Research for Semi-arid Tropics); CRS (Catholic Relief Services): *Seed, Vouchers and Fairs: A Manual for Seed-based Agriculture Recovery in Africa*; and personal correspondence with Oxfam field offices.

4 Bramel and Remington 2005.

Appendix 7

1 Jaspars and Shoham (1999); Taylor and Seaman (2004).

References

Acacia Consultants (2003) 'Evaluation of Project Review of Oxfam GB's 2001–2002 Drought Recovery Programme in Turkana and Wajir Districts of Northern Kenya', Oxford: Oxfam GB.

Acacia Consultants (2005) 'DFID/OXFAM/NOVIB funded NGO Consortium Response to Drought in Togdheer, Sool, Bari / Nugal Regions', External Evaluation Report, July 2005.

Adams L. (2005) 'ODI/UNDP Cash Learning Project Workshop in Aceh, Indonesia', Workshop Report, Humanitarian Policy Group, London: Overseas Development Institute.

Adams L. and E. Kebete (2005) *Breaking the Poverty Cycle: a Case Study of Cash Interventions in Ethiopia*, Background Paper, Humanitarian Policy Group, London: Overseas Development Institute.

Ali, D., T. Fanta, and K. Tilleke (2005) *Cash Relief in a Contested Area. Lessons from Somalia* Network Paper 50, Humanitarian Practice Network, London: Overseas Development Institute.

Bonal E. (2003) 'End of Mission Report – Drought Response Programme, Cambodia', Oxford: Oxfam GB.

Bramel P.J. and T. Remington (2005) 'CRS Seed Vouchers and Fairs, Looking at the effects of their use in Zimbabwe, Ethiopia, and Gambia: a meta-analysis', Catholic Relief Services.

Brocklebank (2005) 'Peer to Peer Evaluation of Cash for Work Programme Carried out in Lamno, Aceh, Indonesia'.

Buchanan-Smith, M. and J. Borton (1999) 'Evaluation of the Wajir Relief Programme 1996–1998', Oxford: Oxfam GB.

Catholic Relief Services (2002) *Seed Vouchers and Fairs: A Manual for Seed-based Agriculture Recovery in Africa*, CRS, ICRISAT, and ODI.

Creti, P. (2005) 'Evaluation of the Livelihood Programme in Mapou and Cape Haitian, Haiti', Oxford: Oxfam GB.

Devereux, S. (1988) 'Entitlements, availability and famine', *Food Policy*, August 1988.

Frize J. (2002) 'Review of Cash for Work Component of the Drought Recovery Programme in Turkana and Wajir Districts', Oxford: Oxfam GB.

GTZ (2005) *Manual of Operations: Pilot Social Cash Transfer Scheme*, Kalomo District, Lusaka, Ministry of Community Development and Social Services / GTZ Social Safety Net Project.

Hadcroft, A. M. (2004) 'Lessons Learned: Humanitarian Response in Haiti – Food Security and Sanitation Component, April-September 2004, Cape Haitian', Oxford: Oxfam GB.

Harvey P. (2005) *Cash and Vouchers in Emergencies*, Discussion Paper, ODI Humanitarian Policy Group, London: Overseas Development Institute.

Hofmann, C.A. (2005) *Cash Transfer Programmes in Afghanistan: a desk review of current policy and practice*, Background Paper, ODI Humanitarian Policy Group, London: Overseas Development Institute.

ICRISAT (2002) 'Organising Seed Fairs in Emergency Situations', India: International Crops Research Institute for the Semi-Arid Tropics.

Jaspars, S. and J. Shoham (1999) 'Targeting the vulnerable; the necessity and feasibility of targeting vulnerable households', *Disasters* 23 (4).

Jones, B. (2004) 'Evaluation of Oxfam GB's ECHO-funded Cash for Work Project in Hazarajat, Afghanistan', Oxford: Oxfam GB.

Khogali, H. (2001) 'Case study: cash for work. Post-flood rehabilitation in Bangladesh', in 'Cash an Alternative to Food Aid', Oxford: Oxfam GB.

Khogali, H. and P. Takhar (2001) 'Evaluation of Oxfam GB Cash for Work Programme, Kitgum/Pader District, Uganda', Oxford: Oxfam GB.

Levine, S. and C. Chastre (2004) *Missing the Point:An Analysis of Food Security Interventions in the Great Lakes*, Humanitarian Practice Network, Network Paper 47.

Lothike, E. (2005) 'Cash: a Currency for Emergency Interventions? Lessons from Recent Experience', presentation made at a meeting hosted by the Overseas Development Institute and the Great Lakes and East Africa Emergency Preparedness Working Group.

Oxfam GB (2002) 'Oxfam (GB) Guiding Principles for Response to Food Crises'

Oxfam GB, TGWU, and Refugee Council (2000) *Token Gestures – the Effects of the Voucher Scheme on Asylum Seekers and Organisations in the UK*, Oxford: Oxford GB.

SDC (Swedish Agency for Development and Cooperation) and UNHCR (2002) 'Compensation for Shelter in Ingushetia'.

Sen, A. (1981) *Poverty and Famines: An Essay on Entitlement and Deprivation*, Oxford: Clarendon Press.

Sperling, L. and C. Longley (2002) 'Beyond seeds and tools: effective support to farmers in emergencies', *Disasters* 26 (4), London: ODI Overseas Development Institute.

Sphere Project (2004) *Humanitarian Charter and Minimum Standards in Disaster Response*, Oxford: Oxfam Publishing.

Taylor, A. and J. Seaman (2004) *Targeting Food Aid in Emergencies*, Special Supplement, Emergency Nutrition Network,

Index

needs assessment 16
NGOs *see* agencies
Niger 72
non-essential items, cash for 9, 13
non-food items
    cash programmes for 30
    see also commodity vouchers

old people *see* vulnerable people
outcome *see* impact

Pakistan 58
parallel economy, risk of 14
pay, in cash for work 53, 60, 62–4
payments
    for cash for work 65–6, 90–1
    of cash grants 39–42, 43
    direct 41–3
    for vouchers 76
Philippines 59
planning
    of cash for work 54–6, 66
    of cash grants 33–4, 41
    of shop-and-voucher schemes 79–80
    of voucher fairs 72–5
process indicators 46, 66, 77, 86–8, 97
project beneficiaries *see* recipients
projects, selection of 54, 57

recipients
    preference for cash transfers 8
    protection from theft 25
    spending patterns 8–10, 13, 14
    see also households; targeting; women
recording
    of direct payments 42–3
    forms 89–91
    of voucher fairs 76
registration *see* targeting
relationships *see* social relationships
risks
    of cash programmes 11–14, 18, 24–5, 32
    of inflation 12, 18, 21, 22
    see also security

Save the Children 9, 40, 63
saving, promotion of 39–40
security 12, 18, 24–5, 43, 50
    see also food security; income security;
        theft
seed fairs 70, 71, 86–7, 88, 93, 97
seed provision, advantages of cash transfers 7
seed vouchers, value of 74
shelter shops 30
shops
    for payment of cash grants 41
    and vouchers 78–82
sick people *see* vulnerable people
social impacts, evaluation of 50
social relationships
    assessment of 17
    see also communities
spending patterns, of recipients 8–10, 13, 14
staffing
    of cash for work 54, 64–5
    of shop-and-voucher schemes 80
    of voucher fairs 77–8
supply chain 92–3
    see also value chain

targeting
    for cash for work 27, 53, 54, 60–2, 95
    for cash grants 35, 37
    for cash transfer 12
    community involvement 12, 25, 34, 60,
        95, 96
    evaluation of 47
    risk minimisation 25
    for vouchers 73
theft, prevention of 25, 43
trade, benefits of cash programmes 10
traders
    encouraging 27, 93
    responses 20–1, 22
    see also shops; voucher fairs
training, for business management 39
tsunami, cash-for-work projects 59–60

value chain 23, 92–3
village relief committees *see* community
  relief committees
voucher fairs 70–2
    community involvement 72, 77–8
    examples 71, 72, 86–7, 88, 93, 97
    management of 75–6, 77–8
    monitoring 77, 97
    planning 72–5
vouchers
    advantages and disadvantages 28,
      78–9
    definition 3
    and dignity 69
    examples 29, 93
    indications for 27, 68
    and inflation 69, 74, 75
    redemption of 76
    in shops 78–82
    targeting for 73
    value of 73–4
    see also cash vouchers; commodity
      vouchers; food vouchers
vulnerable people, cash programmes for
  35, 53, 60–1, 95

wages *see* pay
women
    benefits of cash programmes 10, 12–13
    in cash for work 13, 53, 61
    in community relief committees 34
    income-generation packages 87
    spending patterns 9–10

Zambia 40
Zimbabwe 70, 71, 75, 78, 97